VIRGIN MARY AND THE NEUTRINO

EXPERIMENTAL FUTURES:
TECHNOLOGICAL LIVES, SCIENTIFIC ARTS,
ANTHROPOLOGICAL VOICES

A series edited by Michael M. J. Fischer and Joseph Dumit

VIRGIN MARY

AND THE

NEUTRINO

REALITY IN TROUBLE

ISABELLE STENGERS

Translated by Andrew Goffey

DUKE UNIVERSITY PRESS *Durham and London* 2023

La Vierge et le neutrino: Les scientifiques dans la tourmente
© Les Empêcheurs de Penser en rond /
Éditions du Seuil, 2006
Printed in the United States of America on acid-free paper ∞
Project Editor: Livia Tenzer
Designed by Matthew Tauch
Typeset in Adobe Jenson Pro and Bahnschrift by Westchester
Publishing Services

Library of Congress Cataloging-in-Publication Data
Names: Stengers, Isabelle, author. | Goffey, Andrew, translator.
Title: Virgin Mary and the neutrino : reality in trouble /
Isabelle Stengers ; translated by Andrew Goffey.
Other titles: Vierge et le neutrino. English | Experimental futures.
Description: Durham : Duke University Press, 2023. | Series: Experimental
futures | Includes bibliographical references and index.
Identifiers: LCCN 2023000720 (print)
LCCN 2023000721 (ebook)
ISBN 9781478025207 (paperback)
ISBN 9781478020295 (hardcover)
ISBN 9781478027270 (ebook)
Subjects: LCSH: Science—Philosophy. | Science—Social aspects. |
Feminist theory. | BISAC: SCIENCE / Philosophy & Social Aspects |
SOCIAL SCIENCE / Feminism & Feminist Theory
Classification: LCC Q175.3 .S8413 2023 (print) | LCC Q175.3 (ebook) |
DDC 306.4/5—dc23/eng20230712
LC record available at https://lccn.loc.gov/2023000720
LC ebook record available at https://lccn.loc.gov/2023000721

Cover art: Jitish Kallat, *Untitled (Emergence) Drawing 4*, 2018.
Watercolor, gesso, lacquer, and acrylic epoxy on gesso panel,
40.5 × 30.5 cm. Courtesy of the artist and TEMPLON,
Paris–Brussels–New York. Photo © Adrien Millot.

THIS BOOK RECEIVED A PUBLICATION SUBSIDY FROM DUKE
UNIVERSITY PRESS'S TRANSLATION FUND, A FUND ESTAB-
LISHED BY PRESS AUTHORS WHO DONATED THEIR BOOK
ROYALTIES TO SUPPORT THE TRANSLATION OF SCHOLARLY
BOOKS.

CONTENTS

TRANSLATOR'S PREFACE

ANDREW GOFFEY

———

Virgin Mary and the Neutrino: Reality in Trouble was originally published in French in 2006. As the reader will quickly discover, its starting point is the "science wars" that were raging, especially in the United States, in the 1990s and early 2000s. Such a starting point and the questions that it enables Isabelle Stengers to address were evoked both by the book's original subtitle, "Les scientifiques dans le tourmente" ("Scientists in Trouble"), and by an image on the book's front cover, one of Goya's Black Paintings, *Duel with Cudgels*. But in the years that have intervened since the original text was written, a great deal has happened. The financial crash of 2008; the Arab Spring; the "great acceleration" of climate change; mass-extinction events; the repeated failures, on the part of those Stengers calls "our guardians," to hear the warnings of the Intergovernmental Panel on Climate Change; Black Lives Matter; the COVID-19 pandemic—to name but a few: all point to a situation that is rather different from the one that obtained in the early years of the twenty-first century. The rising tide of mud in which Goya's duelists slug it out seems now to have all but immobilized the scientists, while a sizeable number of the people who might once have been persuaded that this was indeed a spectacle worth watching seem to have turned their backs on science altogether. The squabbles of those who took themselves for heroic agents of Western progress have faded into the background, while denialist purveyors of fake news and alternative truths are mobilizing around a different kind of conflict altogether: it's not just the scientists who are in trouble now.

Changing the title of a book in the passage from one language to another is hardly unusual. The main title of Stengers's first book, written with Léon Chertok, *Le coeur et la raison* became *A Critique of Psychoanalytic Reason*, for example. *Faiseuses d'histoires: Que font les femmes à la pensée*, written with Vinciane Despret, became *Women Who Make a Fuss: The Unfaithful Daughters of Virginia Woolf*; and her book, written with Philippe Pignarre, *La sorcellerie capitaliste: Pratiques de désenvoûtement*, became *Capitalist Sorcery: Breaking the Spell*. While such changes are typically prompted by linguistic difficulties in the translation process, the constraint that this process creates also offers the chance to generate interesting new proximities or effects as a result. In the migration into English, *la Vierge*—with the definite article—has become *Virgin Mary*—still the Blessed Virgin but, more decidedly, she to whom an intimate address is made.[1] For Stengers, the translation process also offers an opportunity for rewriting that operates in a similar way—modest changes to a passage, an extra footnote here or there, sometimes modifications that are a bit more substantial.[2] Shifts and changes of this kind are entirely of a piece with her practice, which is articulated around the careful, experimental posing of problems, an attentiveness to what they demand, and an open acknowledgment of the risks that creative solutions to them could involve. As with Alfred North Whitehead recomposing his *Process and Reality*, it is as if the process of translation gives Stengers, as author-reader, an opportunity to sense the work and the way it functions a little bit differently.[3] A translation might, in any case, be imagined as one of a number of *versions* of a text, in the sense that Stengers's coauthor Vinciane Despret has established for the word *version*,[4] inseparable from the *becoming* of a text registered through the muted qualifications, nuances, and hesitant objections that author-readers might be prompted to make to it in the transition into another language.

Virgin Mary and the Neutrino: Reality in Trouble is not a substantially different book from the French original (notwithstanding the obvious shift from French to English). It's not what Stengers herself calls a *re-articulation*, which she has done with her most recent book, on Whitehead.[5] Some small changes have been made to the text here and there, occasionally picked out in footnotes. But mostly it is the quite considerable interval of time separating the original writing of *La Vierge* from its translation that has imposed the need for this brief preface. The "questioning situation"[6] and Stengers's thinking with regard to it have changed since 2006, and in much of her work since then, it is the challenge that she addresses here—"making the questions that mark an epoch matter, using philosophical means"—that has started to resonate a little differently. Would the book have been written differently, had its author been writing in 2020 or 2021, after the welter of events evoked earlier? Almost certainly. But to say that is not to suggest, implicitly or explicitly, that historical context explains text, as if that might then stand as a reason not to read any further or to suggest that, fifteen to twenty years later, we can now see more clearly what the author, back then, couldn't. And it would also risk missing the specificity of the process that a book such as *Virgin Mary and the Neutrino* engages in. Neither *La Vierge* nor this translation should be read as forming a definitive set of statements concerning the ecology of practices. It is not a "theory" offering an all-terrain account of—and means for judging—a state of affairs, and Stengers is not a theorist (whom she sometimes refers to humorously as "prophets"—authorized by something higher than themselves). But nor is the ecology of practices a static intellectual production. *Virgin Mary* doesn't simply restate in a more condensed form the discussions of Stengers's *Cosmopolitics* (published in French in 1997), which centered on scientific practices. Here the practices in question stretch well beyond science, and the effects aimed at are correspondingly more challenging: how and at what price can we assert that the Virgin Mary, as well as neutrinos, belong to what we call "reality"?

Stengers is a pragmatist, an experimental thinker, for whom thinking *with* and learning *from* is a stance that is unavoidably both speculative *and* practical. Encounters with others matter, and it is perhaps the case that she would not have dared to ask some of the questions that she asks were it not for these ongoing, symbiotic processes. What matters for Stengers, what she does in her work, what her philosophical writing practice embodies, is to engage with certain scientific, technical, and social practices that are *underway*, proposing adventurous, risky explorations, whether those be in Ilya Prigogine's laboratory, in Léon Chertok's persistent but perplexed engagement with hypnosis, in Tobie Nathan's ethnopsychiatric clinic, in neopagan witchcraft, or indeed in other activist practices of "de-enchanting."[7] In some of these cases, one might say the explorations are bound up in controversies internal to a practice, which have not achieved what science studies scholars call closure or become settled matters. This would be the case with Prigogine's work on irreversible processes, perhaps. In other instances, they have generated broader, and sometimes quite fierce, polemics (as with the reception of ethnopsychiatry), in still others, sneering, unconditional dismissal (neopagan witchcraft). Stengers's interest in these practices does not stem especially from a taste for provocation (of the kind that reductionist science seems compelled to engage in, in order to verify that it really is wounding a "humanity" thought subject to credulous illusions).[8] Rather, the practices that Stengers thinks with and the controversies that surround them enable her to raise questions, to formulate problems, and to engage in conceptual experimentation that opens up possibilities. The mode of existence of the physicist she outlines in her *Cosmopolitics*, a "psycho-social type" whose passion for truth does not consign him/her to claim a transcendence for that truth over all others, who might perhaps ask questions that "true" scientists are not supposed to ask, is the product, in part, of this kind of engaged encounter. Stengers refuses the tribunal of judgment and the redundancy and destructiveness of the "we now know" with which that tribunal is associated, and she does so in full and consequential acknowledgment of the pragmatic importance both of events and of others. "Unknowns," in this sense, are constitutive of her philosophical practice.

Underlining the simultaneously speculative and practical dimensions of Stengers's philosophical practice helps when addressing the effectively performative quality of her work—she has referred to the ecology of practices as a "performative ethology of manners of affecting and being affected," for example. The "types" that she fabulates as part of that practice serve a diagnostic function but not without simultaneously generating a possibility for thinking, feeling, and acting differently. Practitioners are offered the possibility of presenting themselves as "idiots"—a type for whom there is something that is "more important," which situates them as minoritarian, following their own line of divergence. To the extent that this happens, they betray the language of consensual evidence that is part of the sleepwalking professional's habitual presentation and may become able to participate in ecological relationships, relationships between diverging ways of making the world matter. The problem of *coherence* with which the ecology of practices communicates, in a modernity riven by the bifurcation of nature (a central theme in her readings of Whitehead), is always a coherence that is *to be created*, one that bets on the possibility that scientists—and others—are capable of doing things a bit differently. In this respect, an ecology of practices is nothing without the catalyzing effects that it aims at (without knowing what such effects would look like) and which, in turn, give it its truth. Stengers refers frequently to Gilles Deleuze's formulation of politics as "needing people to think," not in the sense of confirming the veracity of an analysis but rather in the sense of exploring the possibilities for thinking, feeling, and acting that are implied in the creation of coherence.

Toward the end of *Virgin Mary*, Stengers engages in a discussion of a specific issue that the ecology of practices, she says, must protect itself from. That issue is the seeking of a guarantee with regard to the difference between "what must be taken into account and what one has the right to neglect." One of the reasons for adding the prefix "cosmo-" to politics, she points out, is to remind us that we are not alone in the world, that when we assume the position of being the brains of humanity, the forgetting and humiliation of victims tends to follow. This, I think, offers us a cautionary reminder about not seeing in her conceptualization of the ecology of practices something like a philosophical "system" that might, in fact, be susceptible of a definitive formulation: here at last is the key to the challenges we face, so we now know what we can neglect. The "pharmacological" functioning of the guarantee is something that many decades of the equating of science with progress have perhaps anesthetized us to.[9] But the humiliation that follows when it is assumed one knows what has to be taken into account and what can be neglected points toward an issue that the lapse in time since the original publication of *La Vierge* has rendered especially acute. For humiliation is indeed one of the consequences of what Stengers has, more recently, characterized as the political and cultural disaster of the "defeat" or "undoing" of common sense and its striking capacity for doubting, ruminating, hesitating.

The defeat of common sense, explored in *Réactiver le sens commun: Lecture de Whitehead en temps de debacle* (2020), seems to be most strikingly evident in the contemporary proliferation of fake news, climate-change denial, and conspiracy theories. The refusal to think that we are confronted with in such a proliferation, in turn, provides some with "evidence" that allows them to justify the positions they took in the science wars several decades ago: evidence of the fundamental irrationality of the public ("You see?! People cannot be trusted to think for themselves!"). But there is more to it than this. The defeat of common sense was not something that happened just recently: the sciences' wars—wherein the combatants shared in their exclusion of nonscientist others, with their capacities to object, to hesitate, to laugh, even—unfolded in a landscape that has in some respects been several centuries in the making. The curious invention of bodies that get better for "the wrong reasons" and the history of hypnosis associated with it can be traced back to 1784, for example.[10] And while there are numerous practices that have learned to make sense in common, in the face of the arrogance and ignorance of "our guardians" (who nowadays cannot permit doubts, objections, or concerns, for fear of losing the race to the bottom[11]), those practices nevertheless testify, albeit in a different way, to the same disaster as conspiracy theorists and fake news peddlers, to the same failure on the part of the moderns to exercise due vigilance over their abstractions, to pay attention to what these abstractions allowed them to ignore. Both QAnon and, say, Resilient Indigenous Sisters Engaging (RISE) Coalition, in very different ways, testify to the same catastrophe. But if both are a part of the contemporary situation, Stengers's concern is nevertheless to address what the current epoch might still be capable of, and in this regard, it is because of activists that she feels able to bet that "the somber will not to think, not to be shaken up by anything that might, rightly, scare us, [will] not confront us with what we would have to acknowledge as the sad truth: a common sense that needs to believe in the authority of those who know, because it would be desperately incapable of differentiating between knowledge and opinion."[12]

In this more recent work, Stengers extends the concern with ecological disasters already addressed in *Virgin Mary* and prolonged in different ways in numerous other publications since 2006. But there is, it seems to me, both a difference in tone and a more direct engagement with "epochal" questions that registers the shift in her thinking. The "tentacular" version of Whitehead's metaphysics that she constructs to facilitate the reactivation of common sense, a version that draws on contemporary biology, Harawayian sympoiesis, decolonial anthropology, and feminist and ethnic-minority activism, responds specifically to a "collapse" that raises serious questions about the epoch we find ourselves in. "We do not know" she says, if this epoch "marks the end of modernity or is exploring the possibility of its becoming civilized." The situation that we find ourselves in now is one that creates a "genuine option" in William James's sense[13], the option, as she puts it, of "learning to think without the security of our demonstrations, of consenting to a world that has become intrinsically problematic,"[14] that poses "intrusive" questions and in turn generates doubts, hesitations, and other difficult-to-express concerns. What the place of the old tradition of philosophy and the adventure of ideas will be in this situation is not clear. Indeed, Stengers asks whether "we" still even need philosophy, as we learn to live in the ruins. It not a question that she can answer; it will pertain to the epoch to decide.

This translation was undertaken in difficult circumstances. I'd like to thank Isabelle Stengers and Ken Wissoker for their extraordinary patience, Paul Bains for his helpful comments on an early draft of this translation, and Lynne Pettinger for her unstinting support.

CHAPTER ONE

SCIENTISTS IN TROUBLE

Out of the Frying Pan into the Fire!

It was just over ten years ago, in 1994, that the academic-mediatic tragicomedy, since called the "science wars," erupted in the United States with the publication of *Higher Superstition: The Academic Left and Its Quarrels with Science*, signed by the biologist Paul R. Gross and the mathematician Norman Levitt. Two years later, Alan Sokal's famous hoax relaunched the polemic and led to it crossing the Atlantic. More or less everywhere, scientists mobilized to denounce the "imposters"—philosophers, sociologists, *cultural studies* specialists—who dared to reduce the sciences to social practices like any other, thus promoting a particularly fearful phenomenon: "the rising tide of irrationality" among all of those who constitute what is called "the public."[1]

The first of the absurdities that, for Sokal, should have alerted the editors of the journal *Social Text*, which accepted his hoax article for publication, was the very possibility that a physicist might be susceptible to writing: "It has thus become increasingly apparent that physical 'reality,' no less than social 'reality,' is at bottom a social and linguistic construct; that scientific 'knowledge,' far from being objective, reflects and encodes the dominant ideologies and power relations of the culture that produced it; that the truth claims of science are inherently theory-laden and self-referential; and consequently, that the discourse of the scientific community, for all its undeniable value, cannot assert a privileged epistemological status with respect to counter-hegemonic narratives emanating from dissident or marginalized communities."[2]

If I had been the editor of *Social Text*, I don't know if I would have spotted the other "absurdities" hidden in the proliferating, jargon-saturated details fished out of the literature that it was a matter of disqualifying. But I certainly would have noticed the nonnegotiated shift from the adjective "physical" to the adjective "scientific," and I would have concluded that whatever the goodwill with which this physicist launched into a "cutting and pasting" of more or less interesting arguments, he shared with many of his colleagues the same blind spot: there is no distinction between physics, scientific knowledge, and "science" worth discussing, or even thinking about.

However, this type of detail didn't make the scientists who mobilized after Sokal hesitate. For those whom they attacked, the amplitude of this mobilization confirmed the well-founded nature of their own position. Sokal claimed to have "demonstrated" that the editors of *Social Text* were susceptible to publishing anything. But the manner in which this "demonstration" was transformed into a global denunciation really does seem to demonstrate that what "links" scientists together has something to do with what is often called a "dominant ideology." As for power relations, they really are there, exposed, laid bare, by a reaction as brutal as it was summary.

If the brutality of the arguments mobilized by scientists in the science wars surprised me, I wasn't, however, in the least bit surprised by the conflict itself.[3] This was not just because those who "do science" could not bear to see science "such as it is done" brought to light by "those who study science." Rather it was because of a theme, repeated in a more or less accentuated manner by many of those who study science, that in fact constituted a veritable declaration of war: the sciences are practices just like any other.

However, at the time when the heavily mediatized "science wars" exploded, a process was unfolding silently in the fields of biotechnology and biomedicine that was the bearer of a radical transformation of the craft of the researcher so noisily defended by mobilized scientists. In a not-too-distant future, this transformation might well show that those who endeavored to demonstrate that science is just a social practice, like any other, were indeed quite right.

Progressively, obstinately, the protagonists of industry, finance, and the state, who are the traditional interlocutors of researchers working in the academic milieu, have effectively quit the role that they were supposed to play, breaking the pact that scientists believed they had made with them.[4] The patenting of experimental results suddenly called "inventions," the systematic encouragement of partnerships with industry, big research programs, "spin-offs," and so forth are today announcing what scientists more than a

century ago feared: an enslavement that they believed they had warded off
with the argument of the "goose that lays the golden eggs," the goose who
must be fed without there being any conditions placed on her, so that society can benefit from her eggs. Today, those whom this argument should
have convinced to maintain the appropriate distance are no longer satisfied
with what are called the by-products of academic research, that is to say, of
research that is defined as a function of priorities that researchers themselves determine (between) themselves. They meddle directly with something whose autonomy they were supposed to respect. As a consequence,
the future could easily see the emergence of researchers who will find it
perfectly normal to serve the interests of those who feed their research.
Isn't this already the case for the majority of researchers who are employed
in the private sector?

It is difficult to determine to what extent the scientists mobilized in
the science wars have thought about the link that I have just proposed.
It is probable that they were all the more furious for feeling vulnerable
and threatened, but for the majority, the uncomfortable feeling of no longer being respected "like we were before" and the anger at being insulted
have become confused. Or even that a displaced cause-and-effect relationship provoked them into designating the critics of science as having prime
responsibility for the rising tide of public irrationality, which politicians
would slavishly translate, allowing themselves to be dragged toward the
irreparable: the slaughtering of the goose that lays the golden eggs. It is
also probable that the vast majority of these scientists consider that what
is in the process of being destroyed was a perfectly "normal" distribution of
responsibilities that made for an adequate alliance between the interests
of "science" and those of "society."

However, and this is something to which I will return, what scientists
judge to be "normal" actually constitutes a very particular setup, what
Dominique Pestre has called a "regime of knowledge."[5] The regime of "the
goose that lays the golden eggs" goes back to around the 1870s and doesn't
in the least bit correspond to a logic of the "development of the modern
sciences." What is more, critical thinkers about science are all in agreement
in their description of the lie constituted by the official presentation of the
sciences.[6] In what gets called the linear model, the goose is left alone to
do research as she sees fit; her eggs are golden as a result of the technico-
industrial innovations that flow from her research, and progress, which
justifies the way that the powers of the state feed her, then ensues. This regime of knowledge is instead characterized by a mass of relations of a new

intensity between academic, state, military, and industrial interests. One could thus say that what has now been destroyed is the lie of autonomous sciences presenting themselves as "disinterested."

However, there is nothing automatic about the relationship between the destruction of a lie and whatever might have an interesting relationship with a truth. For my part, I wouldn't consider an effective demystification of the sciences, or a future in which scientists would cease to feel insulted by the thesis according to which their practice is simply a practice like every other practice, to be a victory for critical lucidity. It's not the case that scientists would have been convinced by the arguments of their critics, actually. But they would have been beaten by what sought to bring them to heel, to reduce their activity to a function that plays a role in a "knowledge economy," indeed making them, like everyone else, actors mobilized by the interminable global war of economic competition. It's the destruction of what, in the rest of this book, I will characterize as a "practice," to be sure, but a practice that is not like every other practice, in the sense that no practice is "like every other" practice. As for today's righters of wrongs, if they identified this prospect with the death of the "ideological" illusions entertained by scientists, they would, for their part, be "like all the others," that is to say, like all those who have greeted the destruction of practices "unlike any other" with indifference or irony, because the manner in which such practices sought to defend themselves seemed artificial and mendacious. So much destruction has been celebrated as an elimination of parasitical causes, leaving room for the just and unanimous struggle of humans finally united against the grand destroyer![7]

We owe to Michel Serres a very fine and very somber reading of a painting by Goya that might be appropriate for the "science wars": two men fight to the death while being slowly swallowed by the mud and sand around them.[8] I could add myself to this painting, as I am going to try to intervene in the fight, to create the words that dislocate the confrontation between scientists and their interpreters.[9] The mud that rises will engulf me, as its rising signifies the condemning of my intervention as futile, identified with a defense of privileges belonging to the past, the memory of a time when scientists thought that they could escape the common lot in an enduring way, making themselves respected in a world that doesn't respect much at all.

To whomever would see in the imposition of the common lot a well-deserved punishment I would like to respond that the rising level of mud will not differentiate between us. Where does the practical freedom of those who deploy a historical, sociological, critical, and reflexive analysis of scientific ideology arise from after all, if not the same consensual reasons that scientists take advantage of? Are they not themselves the marginal beneficiaries of these "privileges," which little by little are being dismantled, of this institution of "disinterested" public research whose illusions they denounce? To be sure, what they are doing really is "disinterested," in the sense that they cannot, for their part, talk of the goose that lays the golden eggs: the by-products of their own research can't be patented. But haven't they been tolerated until now because screening them out would have been too costly, that is to say, too conflictual, too noisy, likely to politicize a process that presents itself silently as consensual rationalization? Nothing guarantees that in the short or medium term, the (admittedly dramatically restricted) ground that they benefit from will not crack apart in one go, amid general indifference, because the institutions that accommodate them will suddenly appear to be unjustifiably atypical, reminders of a long-expired past.

The Unexpected Appearing of the Public

To admit that we are, historically speaking, in the same boat, or threatened by the same mud, or defined by the same, now more-or-less moth-eaten, privileges is not easy for the academic world. The tradition there is more usually one of cold war, of reciprocal contempt, and of mutual disqualification. The position of reflexive, critical, even subversive, lucidity is appropriate in certain fields, while in others the position of indefatigable laboring at proof, of the humble artisans of facts that will shut the garrulous up, is appropriate. The glory of the former lies in denouncing the collusion of the latter with power, who will in turn define the lucid subtlety of their critics as sophisticated but empty waffle, pure parasitism, or in need of "real" science. The balance of probabilities is such that they are all incapable of defending the few freedoms that they do enjoy, other than as rights based on consensual reason.

Given the rather sad state of the academic "boat," it is understandable that one might give up defending it and equate instead the destruction of the "moth-eaten old world" of academic privilege with a judgment of truth. That the academic world seems incapable of resisting would be the proof that it was only an artificial construct, the reasons for which were grounded in nothing more than the epoch in which the state and capitalism had asked reason and progress to bless their articulation. And those watching this academic squabble from the sidelines will add that we should rejoice in this change of epoch that sweeps away what was a vector for confusion, clouding the correct point of view. In this case, let's allow the mud to swallow up what was, after all, only a mendacious construct; let's allow the capitalism of today to undo what the capitalism of yesterday had an interest in favoring, or what it asked the state to take responsibility for, in the name of that other mendacious construct: the general interest.

The mud doesn't need our encouragement to rise, but it does benefit from any vulnerability, any identification with a position considered normal, a given, that cannot be called into question, any feeling of security and legitimacy. This, at any rate, is the conviction that has made me write this book. I don't know how the fighters in Goya's painting could resist being swallowed up. I do know that their affronted legitimacy, the rage with which they fight to reestablish a "normal" world, that is to say, a world from which their adversary would have been eliminated, prevents them from asking the question, from risking the possible, where being engulfed is almost certain. I also know that this combat excludes those who, painter included, look on from the sidelines, ironically or fascinated. How to intervene, how to envisage what might be possible, if that signifies interrupting those for whom nothing is more important than managing to finish the other off? The scene excludes the possible as well as third parties.

Betting on another possible history is thus, and in the first place, to expose oneself to condemnation by fighters who are "possessed" by what commands them to make their truth triumph, a truth that needs the defeat of the other. It is to speculate about a transformation of the scene. It's not a matter of dreaming that the fighters would forget about what makes them diverge, of appealing to a unanimous resistance to the rising mud, to a grand reconciliation between the laborer of proof and the subversive critic. Rather it is a matter of repopulating the scene with new protagonists, some interested in what critique is capable of, others by what proof can do, and still others who are interested by aspects of the landscape that the two fighters would agree are of no interest at all, judging them to be secondary.

Speculation is often denounced for its abstract character. To become concrete, the possibility at which it aims does indeed need something that speculation cannot provide. However, speculation can also be stimulated by the sense of a possibility that has already started to be actualized, even if for the time being it has no other power than that of causing a disturbance on the margins of the self-evident. That is the case here because, in a relationship that is strangely contemporaneous with the direct taking in hand of research, another type of process has been set off: the emergence, from within what is globally called "the public" (which is generally either assigned the role of "benefiting from the golden eggs of scientific progress" or of being the dumbfounded spectator of the fight to the death of combatants), of the capacity of quitting this role, that is to say, of learning to meddle with what should not concern those who have been put in the situation of being simply beneficiaries or spectators.

The calling into question of progress by those who were supposed to benefit from it takes many forms, which are sometimes hesitant and stammering and sometimes can confirm the irrationality of opinion in the eyes of the academic world. But the theme of the "rising tide of irrationality," the established manner for judging any calling into question of the "progress" that "science" authorizes, stutters more and more, to the extent that the questions, the objections, and the analyses of those who would contest it become more and more precise and pertinent.

Subversive critics might be tempted to take delight in the gradual loss of power of the grand theme of progress, an essential ingredient in the dominant ideology, one that has enabled any criticism to be kept at a distance or silenced. But it's not their criticisms that have been effective, and what is in the processes of being produced has no need of the demystifying skepticism of academics, who sometimes assert that they represent it. Protesters, those who are concerned by GMOs, by nanotechnology, or by many of the other grand projects that are supposed to resolve humanity's problems, have little interest in the idea that "nothing holds," that everything is an affair of ideology. What they are in the process of learning, and what renders them formidable, is that there is no need to "go back to the origins" of the arguments of the scientists in order to contest their claims. In the case of GMOs, there is no need to "undo" the link established by scientists between nucleotide sequences in a molecule of DNA and the amino acids of a protein. It is enough to follow the arguments made by biologists promoting GMOs to learn to spot the manner in which they "judge," attenuate, or eliminate the differences between "their" laboratory GMOs and what happens or could happen "outside the laboratory," in the field, and within industrial strategies. Protestors have learned that what might hold together in the hands of the scientists stops holding together, or is held together by completely different means, when it changes hands, with scientists then turning into propagandists. Once it is a matter of the master articulation of "scientific progress" and "human progress," the rule of rhetoric begins, with the possible evocation of the idea that all progress has a cost or that the technology of tomorrow will be able to resolve the problems raised by the technical innovations of today.

What we are witnessing today, without which the speculative possibility that I am defending would be empty, is the beginning of a veritable learning process, by means of which "nonscientists" produce the means for meddling with what was supposed not to concern them. It is not a matter for them of meddling with what I would call strictly disciplinary problems, that is to say, with problems that have consequences only for the members of the community for whom they are posed, but of propositions that, in fact, interest our collective future—the very propositions that the scientists' usual protagonists—industry, funding bodies, the different representatives of the state—are also interested in, or which they neglect.

One might say that scientists today are caught in a pincer grip between those who have now stopped respecting the appropriate distances and those who now refuse to be kept at a distance. But there is a crucial difference between what might appear to be the two jaws of the same pair of pincers, and it is on this difference that the possibility I am defending, which has made me write this book, depends. One half of the pincer's jaws profits from the consensual reasons behind which scientists shelter. It takes advantage of a "scientific mind" that scientists use as a vindication so as not to have to worry about what might be "unscientific"—a vindication that after all is fairly well-described by this goose, which honks and lays eggs but which is incapable of defending itself, only of moaning about "irrationality." The other half appeals to scientists who would not only be able to resist, to think, to imagine—they do exist, and numerous are the biologists who have denounced the irresponsibility of sowing GMO seeds, for example— but also to loudly assert that they need the intervention of nonscientists, without whom they would have no chance of being heard. It needs scientists who will stop moaning about irrationality and evoking the painful nostalgia of the heavenly golden age when everyone respected them.

The bet that I am making with regard to the future of scientific practices thus implies this new, still hesitant, process, and its aim is to speak of its importance and its constraints. The idea of a science with "good will" at the service of democracy and the common future seems badly utopian to me. If it were to become the last word, the two halves of the pincer's jaws really would become symmetrical, their only difference concerning the end that science is to service. It was in order to avoid being "instrumentalized" in this way, reduced to being the instruments of ends imposed from elsewhere, that scientists have defended an *identity*, that of the goose that lays the golden eggs or that of the creative sleepwalker who, above all else, must not be woken up. It will be a matter for me of something very different: thinking a *belonging*. Unlike identity, belonging does not define those who belong; rather, it poses the question: *What does this belonging make them capable of?* Other relations, with milieus other than those that fed the goose for its golden eggs, are possible, but these relations have to be created.[10]

This is why, above all else, it will not be a matter for me of describing scientific practices—any more than any other practices—in a mode that would determine what they are. Descriptions of a science that is "good in itself" but perverted by its relations to power, and which should have its freedom to produce reliable knowledge in the service of all restored to it, are as dangerous as those that unveil a science identified with power, taking an active part in an enterprise of "enframing," of submission to calculation and manipulation. It is also as much a matter of avoiding the utopia of sciences that, for better or for worse, would be unlinked from the interest that they provoke, as it is of avoiding the anti-utopia of sciences with an essential link to power. It is a matter of thinking the possibility of linking them differently, of refusing the confidence that scientists demand, and of trusting instead in the possibility of other relations between scientists and their world.

What Are Scientists Capable Of?

If the question posed by belonging is, "What does it make those who belong capable of?" then it is a matter of a question that is properly speaking speculative, because no one has the answer to it, even those it concerns. And it is a serious question, apt to make people hesitate, because any trivialization of the relation of belonging, any reduction to it being "merely a social construction," with its correlate "that can therefore be undone and remade differently," could authorize something that can only end in destruction. The only generality here is that if the question is open, if it has no final answer, but only partial, circumstantial, answers, *these answers can only be produced by those whom the question has put at risk, those who belong.*

Of course, it is still necessary that the question be posed and understood, that it not be rejected a priori, and that is how it differs from the question of identity. When it is a matter of scientists, the name of this identity is often "the scientific mind." The scientific mind needs a very particular milieu, one that respects its needs, that leaves it free to establish the strictest differentiation between what concerns it—and will be called "scientific"—and what scientists are certainly free to participate in, but only as individuals. To ask scientists as such to think about "nonscientific" questions, most notably those raising objections to the "golden eggs" science lays, would be to kill what makes them true scientists.

It is not enough to oppose to this claimed "mind," which can only be nourished by truly scientific questions, the observation that in fact scientists exhibit considerable plasticity. This plasticity is well-known by historians of science, who know that when new state regulations, for instance, force scientists to take into account questions that were previously reputed to be "nonscientific," they adapt without too much difficulty. But such a capacity for adaptation doesn't answer the speculative question of belonging. In this case, the differentiation between "scientific" and "nonscientific" is in effect simply displaced, while the principle according to which "true" scientists don't have the time to spend on questions that divert them from their real mission remains intact.

If the thesis of this book is indeed speculative, the reason is that it doesn't, for its part, stop at the observation concerning the "plasticity" of the scientific mind. Because the question, "What would the belonging of scientists make them capable of" puts into play the possibility of scientists who are capable of different relations with their milieu; it calls into question the veritable right to being exceptional that the "scientific mind" appeals to when it differentiates itself from "opinion," when it defines this opinion as that which it must be protected from.

Characteristically, when, in 2004, just as the knowledge economy was in the process of redefining its horizons, researchers gathered together in a collective thinking about research (*les États généraux de la recherche*), they produced a summary report that insisted on the necessity of the autonomy of science in relation to the expectations . . . of public opinion. In this report they wrote that "citizens expect science to provide solutions to all kinds of social problems: unemployment, the dwindling of oil resources, pollution, cancer. . . . The path that leads to the answer to these questions is not as direct as a programmatic vision of research allows one to believe. . . . Science can only function by elaborating its own questions, protected from the urgency and the distortions inherent in economic and social contingencies."[11]

That the *Le Monde* journalist quoting the report characterizes it as "timorous" marks just how shaky the consensual evidence is, evidence that the scientists trust in. Because the extract in question mobilizes all the rhetorical habits of scientists and does so in a way that verges on the ridiculous. In the first place, there are the "expectations" of citizens, who, it seems, have put their trust in a science that is able to resolve all the significant social problems, unemployment included! In fact, researchers don't want to put them off: quite to the contrary, it seems that they are right to trust science: a path really does seem to lead, albeit indirectly, to the "answer to these questions." What researchers fear is that this expectation might justify a "programmatic vision" that subjects research to the urgency of these problems. In other words, yes, science will provide the answers that will ensure generalized social progress but only on the condition that one not impose problems on it. What is more, "economic and social contingencies"—that is to say, questions that do not arise from "science"—are made inherently responsible for a "distortion." In other words, the progress that science will end up ensuring—if it is left free to elaborate its own questions—will go to the root of problems in a mode that transcends the economic and the social. Here, then, is one example, among many, of the honking of the goose that lays the golden eggs, defending its autonomy in the name of the golden eggs that it might produce.

It is not surprising that, faced with the rising level of mud, scientists can do nothing other than "honk," moan about the irrationality of a society that no longer respects "its" science, and call for research "to be saved." Have they not themselves contributed to the situation whose consequences they fear? In identifying science by means of consensual reasons—the objectivity, the rationality, and the method that are the privilege of the scientific mind—they have made themselves vulnerable to being called upon to produce an objective or rational approach to all the questions that the public or their funders are interested in. They don't have the words for questions that aren't their own and are reduced to honking that these questions are "distorted," that science can only provide answers if it is not subjected to the urgencies that opinion would like to impose on it.

They don't have the words for questions that are not their own. And that is where the (speculative) question about scientists is posed. About scientists who are able to entertain different relations with their milieus, to present both themselves and their questions in a mode that resists any possibility of being transformed into the approach to problems of a common interest that is "at last rational," and that separates them from any privileged link with a progress that those who would benefit from it have to wait for and accept. The sense of the notion of practice that I will develop throughout the pages that follow is to *associate* it with a bond of belonging that scientists know could be destroyed, *while dissociating it* from their claim to an exceptional status, whether this is based on the opposition of the "scientific mind" and opinion or on the irreplaceable role of the sciences in human progress. Of course, this latter argument is aimed at those who are strangers to the "scientific mind" and at those who have to be reminded of the particularity of their interests and of the need for patience if they want to benefit from the satisfying achievements of progress. A little bit like the brain of humanity having to remind its "body" that without it, that body would not be able to enjoy the comfort of life and would still be shivering in damp caves.

Dissociating scientific practices from the "scientific mind" has nothing to do with an operation of epistemological purification. It is to take seriously what scientists knew when they claimed the role of the goose that lays the golden eggs. Not all questions are equal for them. They thus need a certain autonomy in their choice of questions to be respected, which depends on whether they can ask them according to their own constraints. But it is a matter of dissociating this need from the disqualification of other questions as "unscientific," or irrational. That is to say, of dissociating themselves from a milieu that benefited not only from science's "eggs" but also from its capacity to silence, in the name of objectivity, reason and progress, all those who stubbornly ask "bad" questions.

Bruno Latour has proposed a distinction—between *matter of fact* and *matter of concern*—that seems to me to be crucial here. A *matter of fact* is something that is supposed to produce agreement no matter what opinion says or what values it entertains. It's what the argument "think what you like, but the fact is . . ." refers to, and it's what the "scientific mind" opposes to the ramblings of opinion. On the other hand, a *matter of concern*, a matter of a collective preoccupation, demands that all those who are concerned by it be recognized as a legitimate party and that no "fact" be recognized as having the last word. Dissociating the "facts" produced by scientific practices in relation to the "concerns" that are proper to this practice from the "scientific mind" is to imagine the possibility of scientists able to participate with others in the elaboration of collective "matters of concern," scientists who would be the first to say that their contribution needs that of others and to *publicly* call into question any use of the qualifier "scientific" that would silence those who ask or demand to be heard.

The question of knowing "what scientists are capable of" itself belongs to the register of *matters of concern*. It is tempting to make the role played by claims to objectivity and rationality as much an undeniable *matter of fact* as techno-industrial development in the colonial and postcolonial war machine that enslaves the world. The dissociation of scientific practices and the "scientific mind" that I am risking making could then be accused of seeking to restore the innocence of scientists. But the correlate of this temptation is that critics would in turn be presenting themselves as serving a truth the vocation of which is to create agreement. Worse still, a truth that has no need of them to "become a concrete reality." Because the autonomy of science that such critics are devoted to deconstructing is in the process of being destroyed. As for the "scientific mind," it will survive as long as it retains the power to silence, to oppose the rational definition of a problem to all the "rest."

Linking the "role of sciences in our societies" to a *matter of concern* is not to find scientists to be "innocent." The categories of innocence and guilt are terrible shortcuts that are best left to judges. Rather, it is to affirm the feminist cry, "It could be otherwise," regardless of how firmly established the state of affairs it is faced with may seem. To be engaged by this cry is to give up the position of the (neutral) interpreter, authorized by a state of affairs. It is to make a possibility matter, against the authority of probability—here the unstoppable rising of the mud.

Learning to "Speak Well" of the Sciences

To be engaged, but how? I've learned from Donna Haraway that no knowledge should present itself other than as situated, situated by what preoccupies it, to be sure, but also by what it is susceptible of putting into play, by the manner in which its equipment allows it to address a situation. I could certainly present myself as being part of all those who are threatened by the mud, but that wouldn't be enough, because what the mud is in the process of engulfing is not, of course, restricted solely to the places protected by what was called "academic freedom." All those who had reason to think that in the cruel history of the modernization of the world they were on the winning and not the losing side, all those who inscribed the sense of their practices in the prospect of progress, are subject to the same kind of attack, in the name of general mobilization for competitiveness. Everywhere the sense attached to practices has become an obstacle to the necessity of flexibility and in that respect is equivalent to the "superstitions" to which so-called traditional peoples were attached.

It is therefore as a philosopher that I must situate myself, and more precisely as someone who would never have "named" herself a philosopher if she hadn't experienced this practice as a "creation of concepts," as Gilles Deleuze characterized it, at the polar opposite of generalities that are supposed to concern everyone, or of foundations that permit a distribution of rights and duties, sorting and judging. Belonging to philosophy constrains me. It prohibits me from taking a position of transcendence, "above the fray," but also "outside of the epoch." If philosophy has something to do with a creation, this latter doesn't transcend the epoch; it makes the questions by which an epoch is marked matter, using its own (philosophical) means.

However, to make the question of the sciences matter with the means proper to philosophy is also to struggle against the manner in which philosophy is susceptible to identifying itself when it identifies "science," and more precisely, of course, physics, as an "invasive other" with which a relationship must be fixed. This would only be a problem for philosophers—who do scientists take us, and who do we take ourselves, to be?—if the situation didn't have another aspect to it. The manner in which scientists present the sciences, what they identify with science, the consensual reasons to which they have recourse, the arguments that allow them to denounce the "rising tide of irrationality"—all of that is literally saturated with borrowings from philosophers.

It will be said that these borrowings aren't "truly" philosophy—and, effectively, it is a matter of themes *captured* by scientists and transformed into means for their own ends. However, the multiplicity of these borrowings and the ease with which they are made pose a problem. As if, beyond the divergence between what interests philosophers and scientists, there was a form of connivance that united two candidates for the role of the shepherd leading the human flock down the path to safety. For a philosopher, addressing the question of the sciences would thus be to set out the image of a victorious scientific rationality, to which only the philosopher can assign healthy limits, thereby providing the set of ready-to-use considerations that will give "science" an identity compatible with the claims formulated in the name of the "scientific mind." Even denunciation will do. So when Heidegger assimilates scientific truth to enframing, his diagnostic is repeated by scientists, most notably by cognitive theorists: assimilating thinking to calculation is not a hypothesis but something that must be postulated so as to make of the brain a scientific object.

The philosopher is situated here by the sentiment of shame that she feels when faced with the trafficking of "ready-made concepts," the commonplaces that define the rivalrous identities of "science" and "philosophy." However, shame is not enough, because it can lead to disgust, to the decision to have nothing to do with this kind of trafficking. If she can succeed in making people think, it is to the extent that she enters into communication with the possibility of coming into friction with commonplaces, of rendering them perceptible in order to undermine their self-evidence.

That is why the title of this book brings the neutrino, a creature of experiment, and Virgin Mary, to whom the faith of pilgrims is addressed, into proximity.[12] It's not that I consider the Virgin Mary worthy of particular respect, but because the commonplace that, it seems to me, poisons scientists as much as philosophers, making us "modern," presupposes that she is to be dismissed—like all supernatural beings endowed with the capacity to intervene positively and directly in human lives—as belief and superstition.[13]

Escaping from capture by a commonplace like this exposes me to generating the unanimous opposition of everyone this commonplace unites: most notably, the majority of my academic colleagues. It is this decision to expose myself that positively situates me. I will try to construct the position according to which practices of pilgrimage require the existence of the Virgin Mary in a mode that is irreducible to human subjectivity and according to which it is just as important not to insult pilgrims as not to insult physicists. And I am not doing this in the name of tolerance or because one must respect the beliefs of others. If the concepts that it is a matter of creating around the notion of scientific practice result in accepting the "objective existence" of neutrinos and dismissing the Virgin Mary as the result of human subjectivity, they will have failed, because they will have ratified a modern commonplace and will lend themselves to a capture that will mobilize them for "Science."

Coming back to the "science wars," we can say that they unfold entirely on the terrain constituted by this commonplace. Galileo's stroke of genius, we will see, was to capture the theses of philosophical skepticism in order to put everything that testifies to the power of human fiction into the same bag, that of undecidable beliefs, and to affirm the exceptional status of the "new science." That numerous philosophers have tried to get out of the bag into which Galileo put them so that they might constitute valuable, even crucial interlocutors for the sciences, is part of the history of modern philosophy and its sometimes shameful trafficking. But when numerous specialists in *cultural studies* and sociology undertook to suppress the exceptional status of the sciences and put them into the bag that scientists themselves had conceived for everything that wasn't science, when they identified the sciences as "beliefs" just like any other, that was a declaration of war.

We will not be putting the Virgin Mary and the neutrino into "the same bag," because it is the gesture of disqualification itself, science and philosophy's commonplace, which it is a matter of losing the taste for. The risk taken in this book will be to call into question the "bag" itself. It will be a question of seeking to "characterize well" the singularity of the mode of production of knowledges associated with the modern sciences, in a mode that doesn't insult the scientists but that does separate it from the power of judging other practices as relative to the human, all too human, reasons that imprison and divide humans. It will be necessary to learn to "speak well" of the neutrino and of those who are concerned by it but also of learning to "speak well" of other practices, including those that our consensual reasons have disqualified as arising merely from belief.

Gilles Deleuze has written that if to think is to resist, it is a matter of thinking in the presence of[14] the illiterate, alcoholics, dying rats. Here, in order to resist the consensual reasons from which the sciences have benefited but which now make them vulnerable, we will try to think in the presence of other practices, those that have already been destroyed or are now designated as testifying to the arbitrariness proper to human subjectivity.

CHAPTER TWO

THE FORCE OF
EXPERIMENTATION

The Anomaly of the Neutrino

How is one to "speak well" of the sciences? And in the first place, how is one to resist the slope down which one slides, without even thinking about it, when one identifies physics—the name given since the nineteenth century to the science that Galileo called "new"— with a model that can, through approximation, be generalized to all sciences? How is the link between physics and "Science" to be broken, this link commented on by consensual rationality?[1]

When sociologists of science undertook to demonstrate that science was only a social practice like any other, it wasn't ethology or pedagogy that they considered. They knew that in such cases their demonstration would collide with a statement of the sort, "Yes, but in physics . . ." So they had to aim at the head. Likewise if it is a question of taking up a discussion with scientists, of accepting the test of not insulting them, one has to begin with physics. So I will begin with a historical account that belongs to contemporary physics, that of the measurement of the flow of neutrinos emitted by the sun.

Detecting neutrinos is not simple. Raymond Davis Jr., who put together the first detector for the measurement of solar neutrinos in 1967, had set up his apparatus in the bottom of a mine, because to have any chance of detecting an event that would testify to an interaction with a neutrino, it's necessary to eliminate all the other particles from the flow that enters into the detector. It's not that neutrinos are rare: quite the contrary, according to a picturesque estimate, some one hundred billion solar neutrinos pass through one of your thumbnails every second. The problem is that the overwhelming majority of them pass through the nail, then the thumb, then the ground, then . . . without interacting, that is to say, without leaving any traces. Davis knew that at the bottom of the mine, in the reservoir of perchloroethylene he installed, only ten or so chlorine atoms a week would be transformed into atoms of radioactive argon following an interaction with a neutrino. And they would have to be extracted, a labor worse than looking for a needle in a haystack. One can really talk about detection "at the limits" here, and in any case, that was what Davis was passionate about. He continued to perfect his experimental procedures while the theorists argued—because Davis's measurements resulted in what physicists designate as an "anomaly": the flow of neutrinos that he measured was broadly, and stubbornly remained, smaller than the theoretical prediction that made this measurement interesting.

Davis's passion matters. An anomaly is not a simple deviation from the predictions; it is an obstinate one, implying an effort at experimenting that is equally obstinate, and recognized as such, that is, as resisting any technical "perfecting" by the research community. While the theorists might initially have hoped that the conflict between the measurement and what theory predicted resulted from a weakness in the sensitivity of the detector, over the course of the 1970s, they would acknowledge that the process of detection was robust and reliable. To call it into question could no longer be anything other than to look for a loophole: there really was a discrepancy between experiment and theory.

In 1986, the British sociologist of science Trevor Pinch published *Confronting Nature: The Sociology of Solar-Neutrino Detection*. This book conveys both the strength and the interest of the new sociology of science that was born in Great Britain. It is entirely representative of the rich and passionate work undertaken within this field. But it is also representative of the style that made the scientists angry, because they recognized the mark, if not of a refusal, at least of a fundamental reticence when faced with what matters most to them.

Pinch recounts the history that led to Davis's experiment, as well as what followed, in a mode that was powerful in helping to get rid of the taste for epistemological vignettes based on a simple, logical relation between "fact" and "theory." There is nothing simple about the relationship between a theoretical prediction and an experiment. In order for Davis's experiment to be envisaged—including from a financial point of view—it was necessary for it to stimulate the interest of specialists in three different fields: those who deal with nuclear chemistry (with the nuclear reactions that produce neutrinos in the sun and those that allow for their detection); those who study the sun itself (most notably its abundance of different constituent elements); and those who deal with neutrinos. The very possibility that Davis could carry out his—costly—experiment thus conveys the success of a social construction: legitimate representatives of all three fields came to an agreement to claim that the measurement of solar neutrinos was of the greatest importance. This experiment, they affirmed, "must" be carried out, because it would provide the keystone for the convergence of the three fields, already realized on paper in a model produced by the astrophysicist John Bahcall.

Pinch shows clearly that what the funding application presented as a "theoretical deduction to be confirmed experimentally" translates a montage negotiated between protagonists, each of whom expected repercussions that would be beneficial for their respective fields. However, there is something very particular here about what Pinch calls a "social construction": the negotiation process rendered the convergence of human interests and the convergence realized in the model of the sun constructed by Bahcall inseparable. Taken in isolation, none of the three fields brought together had any verifiable consequence regarding the quantity of neutrinos emitted by the Sun. The Sun, as the source of emissions, was at stake. Only a model that articulated the different theories around the Sun could give these the power to make a difference with regard to solar neutrinos and to predict the number of them that Davis should have been able to detect.

The epistemological vignette, which predicts the experimental confirmation of "a" theory, is effectively undermined here, but not at all in the sense that what would replace it would be an agreement that was "social" in the sense of "only social." The protagonists in the agreement are linked together not by their goodwill, but . . . by the Sun. One might perhaps object that it is a representation of the Sun, thereby bringing the human back into the middle of the picture. But if it is indeed not a matter of the "Sun in itself," nor is it a matter of a "human named Bahcall's representation," either. How can the human named Bahcall and the nonhuman named Sun be opposed, when what organizes the scene is something new, the entering of the Sun into a new connection, which can be phrased "if we accept this mode of characterization, then we should be able to . . ."?

The philosopher William James gave a definition of the "truth of an idea" that pragmatism has often been denounced for: the truth is nothing other than the consequences of this idea, its "cash-value," meaning the interest of the differences that it enables to be made. Now, James's pragmatism is perfectly appropriate, and it takes on all its demanding scope here. It is not a matter of a "calculation" aiming at the most favorable consequences but of an "art of consequences" that accepts that a proposition may be put at risk by its consequences.[2] The truth of Bahcall's model doesn't derive only from its interest for the protagonists in the agreement, in the usual sense of the term, that is to say, the possibility of obtaining the funds for Davis's experiment and of benefiting from the prestige that its success would create. This model also means the possible creation of a new relationship, to be verified, with the Sun. The Sun is now able to intervene in human knowledge in a new mode, to make a new type of difference and to entail new consequences. Or to disappoint those who would try to make it intervene. What "links" the protagonists is now a pact between humans, but a pact that deliberately exposes itself to verification of the manner in which Bahcall's model mobilizes the Sun.

Pinch devoted himself to showing that the reaction of the protagonists to Davis's measurements had nothing to do with an epistemological vignette staging confirmation and refutation. There's nothing surprising about this, because the experimental measurement didn't "refute" a theory but a composite ensemble, and the measuring procedure itself was possible only by putting to work part of this ensemble (thanks to what trace can one identify a neutrino?) If, on the one hand, verification took place, both the components and the model of the Sun that put them together would have been considered to be verified. On the other hand, refutation here is powerless to refute anything, because the protagonists could not agree on what had been refuted. The effort at convergence failed, and moreover, what enabled the composite ensemble to hold together, the model of the Sun, which had been presented as solid, became the weakest link, held responsible for the failure by numerous researchers. It had been necessary so as to bring about an agreement between fields verified through separate experiments, but it was ultimately nothing but a model based solely on indirect inferences. Down with the model!

In 1976, John Bahcall described his sad situation in this way: "Almost every theoretical physicist believes that we astrophysicists have just messed it up and it's our fault and we never understood what was happening in the center of the Sun no matter how much we pretended to do so."[3]

The case of solar neutrinos, as Pinch relates it, is presented according to the rules of the new sociology of science. It aims to show that one can follow actors in the situation of constructing and negotiating a controversy all the better for not letting the "reality" that intervenes only retroactively, after the agreement, align them: those who were right, those who were wrong, and the reasons why they were wrong. Indeed, in 1983, none of the actors can make reality intervene as an argument. Even Karl Popper's notion of the contradiction between theory and fact doesn't hold water. Depending on actor and moment in time, there was a hesitation as to a possible accommodation or the recognition of a contradiction between the model and the measurement. Pinch makes use of these hesitations with regard to traditional epistemological concepts so as to affirm the socially constructed characteristic of what is usually attributed to nature or to logic. The history of solar neutrinos reveals that neither the one nor the other imposed anything on the protagonists.

However, a physicist—Bahcall or Davis, for example—would no doubt have been scandalized if he had read Pinch's book, and all the more so for having to acknowledge that nothing of what the sociologist says is "false." The scandal is that he says it badly.

This "speaking badly," which is not speaking "falsely," but rather speaking in a mode that insults those of whom one speaks, cannot be reduced to a general question of civility. The sociologist might take refuge behind the image of his science as legitimately inflicting a "narcissistic wound" on those it is concerned with. Isn't that what most sociologists go in for? Why hesitate to unveil, to deconstruct, to show what is dissimulated behind the reasons constructed by the actors involved, just because it is a matter of scientists?

We will come back to this conception of the profession of the sociologist. What matters here is how physicists would respond, or rather could respond, if they didn't fall back on the slogan[4] "the rising tide of irrationality!" What they might say to Pinch is that he is not wrong to criticize the vignettes of epistemology, which are of course only valid once the hesitation has ceased. But what is intolerable is that he simply takes the opposing position: as scientific judgment is imposed neither by nature nor logic, this judgment must therefore be "a social construction." As a consequence, what has been classified as the "solar neutrino anomaly" would be explained as a mere failed social construction: in this instance it turns out that the protagonists did not succeed in coming to an agreement about the signification to confer on the results. But the scandalized physicists will protest that protagonists not succeeding in coming to an agreement is not a failure. If there is a failure, it is precisely that no solution *having the power to make them agree* imposed itself. They are therefore within their rights to turn the argument around: if they had come to an agreement about an ad hoc modification of the model, if they had silenced the anomaly, then their agreement could have been reducible to a "social construction"!

What Insults Scientists

If I have chosen the history of solar neutrinos, I have done so because the scandalized physicists would nonetheless have been interested by Trevor Pinch's narrative. He recounts an impassioned story; he creates an appetite for what a "scientific culture" nourished by such investigations would be. It is undeniable that one of the great contributions of the sociology of the sciences has been to reconstitute in great detail the hesitant history of the construction of experimental apparatuses and the facts they generate, which are often forgotten, as if the facts were born "fully formed."

However, a rather curious assertion is often presented, or is insinuated, in investigations such as that of Pinch. Giving their due to the human labor of perfecting an apparatus, the unforeseen turns, the negotiations, the alliances with corporations and public bodies, and so forth, entails dismissing as scientific folklore what it is that matters for the scientists, that is to say, the reference to a "reality." As if humans were the only ones in charge, as if the experimental apparatus, such as it was developed in the end, was the only thing that was really responsible for the "facts" that it produces.

Some sociologists will certainly admit that not everything is possible, that "reality" imposes certain constraints. But it wouldn't be any less insulting, and the scientists would be within their rights to protest that the sociologists are speaking "badly" of what they do: as if the experimenters simply had to accept these constraints as limits on their room for maneuver. They will protest that the laborious modifications of their apparatuses are not "erratic," merely seeking to stabilize their effective functioning around the "desired functioning." They will claim that such modifications must be understandable as perfecting the functioning of the apparatus or as able to be legitimated by other, independent confirmations. To which the sociologist will always be able to reply with a sneer: what "you" call perfecting the apparatus, what "you" define as a confirmation.

This sneering is not intended to be malicious but professional: a true sociologist is not supposed to share the actor's reasons; he is not to become a physicist or a biologist; and he must manage to avoid being swayed by these reasons. In this instance, he claims the freedom to continue being able to say, "It is not *the* representation of the phenomenon; it is *your* representation." It's no doubt because of the power of the scientists' reasoning that sociologists have claimed this freedom, because it protects them from being captured by the authority of those they have taken as objects of their study, and therefore allows them to maintain the kind of distance that is appropriate to a scientist questioning his object of study. If they were studying the activity of dressmakers, for example, they would never have thought to loudly assert that the garment being fabricated owed nothing to the fabric and everything to the human. They would have listened with great interest to a dressmaker speaking about their know-how, of the fold that is created when their fingers meet a fabric the subtle properties of which are not limits but resources.

However, some sociologists, including Pinch, have judged that they are authorized to go a bit further and to announce a highly polemical conclusion: since neither nature nor logic have the power to provide scientists with their reasons, contrary to what the vignettes of epistemology claim, there is something else that does have this power. The link between "reality" and the "power to explain" thus finds itself prolonged in a rather uncontrolled manner: it is social, economic, political, institutional reality that has primacy over the reality invoked by scientists.

Curiously enough, even the authors of the kind of vignettes that the new sociologists and historians of the sciences batter about were sometimes more subtle. Thus, since *The Logic of Scientific Discovery* (1959), Karl Popper was very clear that refutation, in the logical sense of the term, is not imposed on scientists; logic is not the method of the sciences. What singularizes the sciences, by contrast, is the deliberate choice, not constrained by logic, to make count as a refutation what could yet be "accommodated" by an ad hoc modification. Of course, it was the individual scientist, deciding heroically to expose his theory to refutation, whom Popper was putting on stage here, whereas the new sociologists of science have quite rightly emphasized the collective, rather than the individual, characteristic of negotiations as to what does and doesn't count. Thus they have been quite right to reject Popper's scientist as a fiction staging an ethical heroism that they have never encountered. They've never seen a scientist deliberately "choosing" to expose his cherished theory to refutation. In any case, that wasn't what the physicists were after with their solar neutrinos: it was a matter of verifying, and the "refutation" left them in complete disarray.

"Nature does not speak," the sociologists of the sciences continue to affirm. When one observes laboratories, it is the humans who are active, not the phenomena that they study. It is therefore they, the scientists, who are the origin of the knowledges that they produced. However, the question that might really matter is not "who" is active but "how" these particular humans called physicists are active. It was the manner in which they are active that constrained them, in 1986, to recognize that they were unable to resolve the solar neutrino anomaly. The very fact of having recognized an "anomaly," that they didn't use logic, seemingly so accommodating, as a means to sort the problem out, that they hesitated, then becomes a testimony to them knowing that, in this case, effectively, "nature *did not speak*." This also implies that sometimes it really does happen that one can succeed in making nature speak.

Between hesitation, agreement, and success, two rival versions of a narrative can thus be compared.[5] Either the scientists hesitated because, in this instance, they hadn't succeeded *socially* in coming to an agreement on an acceptable mode of accommodation. Or the scientists hesitated because, in this instance, they hadn't succeeded in creating an experimental situation that forced them to come to an agreement that could have made them say, "Nature has spoken!"

Between the two versions, the difference is not that of a thesis that bears on "reality" or on "nature" or on "knowledge" but on the manner of describing the hesitation, the uncertainty, of scientists getting to grips with solar neutrinos. The first describes them as seeking an agreement "among humans" and doesn't recognize any role for the nature that, subsequently, it will be claimed "made" a difference. Whatever the scientists claim, sociological judgment will be unwavering: "Activate yourselves, my friends, imagine, make objections, search for, but know that however you interpret your own activity, for us it will always be *your* activity, and we will always be there to say, "*It is you who are responsible; you made this difference, the responsibility for which you attribute to nature.*" As for the second version, it doesn't describe the scientists as if they were neutral, restricting themselves to recording nature's verdict, but as "obliged." It doesn't claim that "nature" suddenly intervenes and puts an end to hesitations. It maintains that the manner in which the scientists hesitate, imagine, make objections, in short, do research, *testifies to what obliges them*: the possibility that "nature" might make a difference between the different interpretations that concern it.

What the first version opposes—the labor of humans and the belief that they make reality speak—is articulated by the second version under the sign of trust. Scientists would not be activated in this manner if they didn't trust in a possibility, a possibility without any guarantees, to be sure, but which is about something they know "can happen." Indeed their laboratories are populated with instruments that testify to a past where, in effect, it did happen. It can happen that a situation is created that confers on what the question is addressed to the power to make a difference between the different interpretations. That is the very definition of a successful experiment: scientists have succeeded in ceasing to be "free" to interpret as they see fit. *The possibility of this success is what obliges them*, and what makes them consider the first version of the narrative, according to which their free interpretations are and will remain free (even if they give up this freedom in order to be able to come to some agreement), to be an insult.

The second version denies nothing of what sociologists are interested in, but it makes perceptible just how foreign the rivalry that they stage, between two possible causes, two sources of responsibility for scientific knowledges—reality or humans (with their interests, instruments, allies, negotiations, etc.), is to the practice of the scientists they describe. Everything that sociologists attribute importance to matters, of course, but the alternative between rival powers that they invoke abstracts out what makes them a social group, to be sure, but "not like any other." Practitioners hesitate around the neutrino anomaly, because their practice obliges them to hesitate. It obliges them not to decide for as long as they have been unable to create the happy situation in which they will be able to affirm that "reality" was responsible, that it at last made the difference the absence of which had kept them in suspense for so long.

Sociologists of the sciences and other skeptics are evidently right. To constitute "reality" as a "cause," in the sense that knowledge would be produced as its "effect," is more than a caricature: it is a confusion of registers. "Reality" is indeed the "cause" but in an entirely different register, a register in which one can talk of *a cause in so far as it obliges*. What I am going to claim, thanks to the notion of *practice*, is that if "reality" is not the cause of the success of scientists, it is the cause that gives their success its specificity; it is what forces them to hesitate in a mode that diverges from the modes of hesitation of lawyers,[6] politicians, or musicians. The practice of experimenters confers on what they interrogate the power of a "cause" that obliges them to think.[7]

Successful experimentation is not that "nature responds." An apparatus always obtains responses. Thus Ray Davis's detector obtained a response bearing on the quantity of neutrinos emitted by the Sun. But not all responses are equal. What matters to experimenters is, rather, to be able to affirm that here "it really was nature that responded," that is to say, that here no one should be able to transform the fact that an apparatus always obtains responses into an objection or a skeptical shrug of the shoulders. The primary role of "reality," such as it is put in play by the kind of realism associated with successful experiments, isn't that of what responds but of what *vouches for the responses*, of what can be called a "guarantor,"[8] that which agrees to offer a guarantee. This guarantee bears first on the relevance of the question that the experimental apparatus allows to be asked. If the flow of neutrinos detected by Davis had responded to what the theorists anticipated, it would have been possible to affirm that the Sun "agreed" to confirm the well-foundedness of the model proposed by Bahcall, that is to say, to play the role that scientists propose for it in a mode which confirms that it has been addressed in an appropriate manner.

That "reality" can in certain cases effectively "vouch for the response," that is to say, to accept the role that experimenters hope it will endorse, makes these cases "successful." Reality is therefore present through the trust that scientists maintain that they might be able to succeed. That scientists maintain but that obligates them also. Because they can only hope to succeed in making what they interrogate play the role of guarantor if the difference between success and accommodation is for them a *matter of concern*, something that forces them to hesitate, that because of which they hesitate, discuss, imagine, get busy about it, or remain in suspense.

That is why the practitioners of the experimental sciences will save a benign and pitying shrug of the shoulders, or a furious protest, for those who interpret statements authorized by a success in terms of a "justified belief" that is no different in nature from the ancient Greek belief in gods and goddesses, for example.[9] That is also why it will not be enough for them that one acknowledges that they are not free to envisage just any interpretation, that "nature" or "reality" offers certain constraints. As long as these constraints are seen as limiting the freedom to construct, the scientist will remain scandalized, because what is misunderstood is that the aim of the work of the scientists is to succeed in *conferring on what the scientist interrogates the power to constrain him or her*. To be constrained is not a limit but the aim that orients his or her activity, which constitutes its stakes as such. The "happy ending" of a scientific history is the moment when the scientists can finally proclaim that they are constrained to interpret a situation in this way and no other.

Sometimes There Is Dancing in the Laboratories

Let us come back to the history of solar neutrinos, because its punchline constitutes a very typical "happy ending" in science, a success in which the role of guarantor can be conferred on "nature."

Trevor Pinch's book already signals one possibility; Davis's results and Bahcall's model would no longer be contradictory if it was allowed that neutrinos "oscillate" between the three distinct kinds identified by physicists. Everything could then be explained: Davis's detector was able only to detect one of the three kinds, the neutrinos that nuclear chemistry affirms are emitted by the Sun; if one allowed that, once emitted, these neutrinos are capable of metamorphosing, of passing from one kind to another during their trajectory, Davis's mode of detection is no longer adequate. But to allow this possibility, the neutrino must cease to be the particle that physicists define as not having any mass; theory demands that if it is capable of oscillating, it must have a mass, however weak this mass may be.

Why was a hypothesis that would have sorted everything out not allowed from the outset? Here, too, one can propose a "social" answer, that is to say, an answer in terms of relations of force: an experiment involving something as badly defined as the Sun has no right to intervene directly in "big" physics, the physics it behooves to define particles. However, what a sociologist will call a (purely social) relation of force can often also be called a "question of reliability."

One might recall here the history of "water memory" and the indignation displayed by Jacques Benveniste, because he had found that his experiments, which answered to the most demanding criteria in effect in immunology, were refused the power to call into question the manner in which physicists and chemists characterize water. Aren't all "facts" equal? Evidently they are, in the vignettes of epistemology. But certainly not, if one takes into account the opaque, uncertain, and almost inaudible testimony of the living cells that Benveniste asked to play the role of detector of an effect called on to overturn physics and chemistry. This doesn't signify that Benveniste's hypothesis was "false" but that the contrast between "false" and "verified" depends on the labor of verification, which passes via the transformation of an initial, precarious, circumstantial fact into a reliable "reproducible fact." What is more, when Benveniste protested by explaining that other laboratories were not repeating his experiment in an identical manner, he became a victim of the vignette with which he had armed himself, according to which "my fact refutes your theories." He didn't want to acknowledge that a reproducible fact is not an identically repeatable fact but a fact that it has been possible to stabilize the conditions of production and identify the scope of: that of the variants "that work" and those that do not, as well as the reasons, which also have to be verified, for this difference.

Let's come back to the neutrinos and to an anomaly that, thanks to the relentless work of Ray Davis, it was possible to characterize as reproducible, as resisting every attempt at calling into question, every modification of the original detector. Why, it will be asked, was an experiment not imagined that might have tested the hypothesis according to which the neutrino itself was in question? It is useful here to recall Pinch's description linking the (expensive) experiment by Davis with the bringing into convergence of the different fields that the specialists who agreed it was needed came from. How is one to demand an experiment that is even more expensive and that has no other justification than to verify whether, hypothetically, the failure of the previous experiment could not be explained by the neutrinos rather than by the manner in which the sun was modeled? Unlike a police investigation, in which the investigators must "close down all avenues of enquiry," testing and eliminating them as quickly as possible, the physics of neutrinos forms a part of those sciences in which success must now be promised in order to obtain the necessary funds. The argument that one does it "just to find out" is allowed only if the experiment is relatively cheap. For years, then, the neutrino anomaly lay dormant. No one knew how it would one day be resolved: would it be through a modification of Bahcall's model, integrating some unknown peculiarity of the Sun or by a staggering innovation worthy of the Nobel Prize?

Today, neutrinos have a mass, and in 2002, Ray Davis shared the Nobel Prize with Masatoshi Koshiba, whose experiments set the question off again, and with Riccardo Giacconi, who conceptualized an apparatus enabling the source of X-rays from outside the solar system to be detected. This third party indicates clearly that the thesis about "social construction" has not been refuted. Bahcall's work has been validated, but Bahcall has not been crowned. His model of the Sun was "good science," but the Sun is henceforth no longer the issue (with the "henceforth" that can be used once the success has been achieved). Now that "nature has spoken," now that the anomaly has become actively involved in a "happy scientific history," the Sun is nothing more than a first case and an astronomy of neutrinos—that is to say, new observations of stars that are henceforth identified as sources of neutrinos can be envisaged. But to do that, it will be necessary to obtain a great deal of money. And perhaps the decision to crown a pioneer in the astronomy of X-rays with the Nobel Prize will prepare the ground, conferring its prestige on those who will be soon be applying to public-funding bodies in order to place neutrinos in the service of astronomy.

It is therefore not a matter of opposing "social construction" and "true science" but of taking into account that the social character of the success does not contradict the fact that we are dealing with a "story with a happy ending," a history that even includes the "chance discovery" element dear to amateurs. It was the happy ending that was dramatized in the BBC broadcast, "Project Poltergeist," for the well-known documentary series *Horizon*.[10] The scientists interviewed were now in their element: what sociologists of the sciences judge to be impossible had happened, and in the very mode that makes experimenters work, imagine, and hope. "Nature had spoken."

David Wark (who was responsible for the 1999 experiment), testified that "to have spent such a big part of your life on something, which then works, and doesn't just work, but works so marvelously, is the most extraordinary experience that a scientist can have. And what then strikes you is that what you have done really is to learn something about the universe that no one previously knew, and now you can say it!"

What happened? Why did a mega-experiment, one that involved three kinds of detectors and gave the solar neutrino case its final punctuation, become possible in 1999? The reason was that in the intervening time, the Sun was taken "out of" the debate by the work of the Japanese physicists who—and this was by chance—asked a question that their own research program had not foreseen. They were having to deal with the source of a nuisance, a parasitical noise, neutrinos that came from an upper layer of the atmosphere, and they realized that these latter were themselves also less numerous than the theory predicted. One anomaly did not justify an expensive research program. But two independent and converging anomalies did, insofar as neutrinos were the only common point between them. Consequently they legitimated the long and expensive research effort that would finally succeed in constraining the scientists to come to an agreement about these anomalies.

The Japanese team would change its program and construct a giant detector able to identify the direction of travel of the atmospheric neutrinos it detected: did they come from above or from below, after having passed through the Earth? If neutrinos oscillate during their trajectory, there would have to be a difference between the flows of neutrinos detected, according to whether they came from above or from below, as the trajectory of the first would be shorter than the second. And the difference was there! This time the measurement of three kinds of solar neutrinos acquired the status of a key finding, called out for by all the specialists. It happened in 1999. Neutrinos thereby acquired a mass; the Nobel Prize was in view; and . . . Bahcall's model was officially exonerated!

In 2004 Bahcall recalled, "I was called right after the announcement was made by someone from the *New York Times* and asked how I felt. And without thinking I said, 'I feel like dancing, I'm so happy!' . . . You know, it was like for three decades people had been pointing at this guy and saying this is the guy that wrongly calculated the flux of neutrinos from the Sun. And suddenly that wasn't so, and it was like a person who had been sentenced for some heinous crime and then a DNA test is made and it is found that he isn't guilty. And that's exactly the way I felt."

For Bahcall, it was effectively "over" because the success "only" absolved him of the suspicion that had weighed on his life. But one could also say "dancing in the laboratory," what marks the experiment as having a "happy ending," does not mark the end of the story but, rather, a new beginning. Ernest Rutherford danced after the experiment that substituted radioactivity as the consequence of the transmutation of a chemical element into another chemical element for radioactivity as a property of radioactive elements. Irène Joliot-Curie and Frédéric Joliot danced when they succeeded in establishing the existence of artificial radioactivity. In cases of this kind, those who dance know that the event that anchors the trust of the experimenters happened, and they also know that a new chapter will open up, in which new questions can "now" be envisaged and that the history of their science will be punctuated by a "starting from" that they will be part of and that maybe will even bear their names.

The joy of scientists puts those who describe them up against the wall. Should they take seriously the joy of those they are describing, the meaning they give to their success, their delight because they really have "learned" something about the "universe"? The risk that I would associate with the idea of the "social construction of knowledge" is that it cannot take this joy seriously: it is the white lab coat that makes the scientist, and differentiating between Rutherford and an economist going through his statistics would amount to allowing oneself to be captured by the "reality" of the scientists. From the point of view of the practices that I am defending, the joy of the scientists counts. There are laboratories where no one dances, where "it works" doesn't make people dance, or where no "it works" is in play.[11] Success, if success there is, will have to be characterized differently.

However, a problem insists. As has been seen, this characterization of experimental practice doesn't exclude the "social" dimension of this activity. However, the problem is that I seem to be saying, purely and simply, that the scientist who says this dimension is only a secondary detail that in no way affects the fact that "we now know" (that neutrinos have a mass) is correct. It is indeed here that the danger resides, here that one must slow down. Does acknowledging the successes of scientists mean admitting that the sciences, or certain sciences, really can attain reality beyond the "social," in other words, attain a reality that is "independent of humans"?

It's necessary to slow down, because we are dealing with something that could give all its meaning to the "first version," that of the new sociologists. Its primary vocation might not be to describe scientific practices well but to describe them in a mode that creates an obstacle to the claims of scientists, coming out of the laboratory in which they have been dancing, to announce to the public that what they have discovered concerns all human beings because it is a matter of a reality that transcends opinions, cultures, political convictions. In other words, the sociologists would be refusing the "objective facts" of their physicist colleagues because they know that these facts will be presented as "valid for all." Whatever humans' values or culture might be, they will have to kneel "before the facts," like the scientists themselves, who are the first to do it.

Has Reality Spoken?

We will have to revisit the manner in which this devastating claim has been constructed. But before doing that, one must accept the question and pause on the fact that the success of the sciences, or at least, of those sciences in which experimental success can make people dance, can be presented as the discovery of an "independent reality" that ought, as such, to concern all human beings.

This mode of presentation of the success of experimenters institutes a situation of confrontation, not just with sociologists but with all those knowledges and practices that by contrast are reputedly "not objective," that "mix up" facts and values. This, moreover, is the reason why some sociologists have transformed their demystificatory undertaking into a democratic crusade, making the generality of their categories communicate directly with defending the legitimacy of opinions in the face of a knowledge that claims to transcend it: *we will show you that this statue, which claims to dominate you, has feet of clay!*

However, it is the very image of the statue that poses a problem, and not its solidity. Recall here that the possibility of attributing a mass to neutrinos does not deny any "democratic" opinion, and it does not "dominate" anybody. It's the same for the link between sequences of DNA and sequences of protein. All the terms in play here—sequence, DNA, protein—were born in the laboratory, and no one outside had an "opinion" about them that the verdict of science would happen to contradict. Scientists have succeeded in producing a situation that *permits them* to "kneel before the facts," and that is what made them dance, but it is their own "opinions" that the facts have acquired the power to arbitrate between. Why would humanity as a whole suddenly have to bend its knee—in an entirely different mode, that of *grieving for* something that mattered but could divide—by recognizing that experimental success permits a reality that transcends humans to be designated?

One rightly feels that there is something bizarre about this transformation. Experimental success would not have been possible if it hadn't mattered for those it brings together, leaving everything else in abeyance, yet suddenly it permits something to be affirmed that now has to matter for humanity as a whole. The word *importance* has surreptitiously changed its role: it went from designating the singularity of experimental practice to designating a "reality" that is supposed to be everyone's.

The word *reality* has also changed meaning. When it makes scientists dance in their laboratories, the independence of reality is celebrated because this independence is the condition for the event: they organized their experiment as a rendezvous where "reality" would show up if and only if the experimental design was relevant—thus endorsing the role of "guarantor," corroborating the relevance of the question it would respond to. Or accepting to enter into the offered relationship. Even if the formulation is a bit on the long side, it fits perfectly with the happy ending that I have just described. When the Japanese physicists brought into the picture neutrinos that had traveled different distances, I was able to write, "The difference was there," and obviously it wasn't about a "rendezvous" in the usual sense, implying that what shows up does it intentionally. It was a matter of celebrating the independence that matters to the physicist, the fact that her success depends on something that she cannot command. But I did not write that experimental success gives access to a reality that is "knowable" as such, independently of the experimental question.

Responding to the relevance of a question does not signify being defined by this question. To neglect this distinction is to allow oneself to be captured by "grand" alternatives with a philosophical allure, in which it is a question of a reality that is "knowable" "independently of the observer or not," or in which experimental success finds itself linked to questions posed by "knowledge" in general. And where, fascinated by these questions, one forgets that the success of a rendezvous does not allow what shows up to be known "as it is," but only to confirm the hypothesis in terms of which the rendezvous has been organized.

If a philosopher asserts, in the manner inaugurated by Kant, that phenomena are predefined according to principles that are also the categories used by our questions and can therefore only confirm the well-foundedness of these categories, the experimenter will shrug her shoulders and move on. Or she will stop and will ask the philosopher if Kant really did mean that the categories of the understanding allowed the neutrino question, with or without mass, to be settled. By contrast, she will not shrug her shoulders if she is suspected of having produced what is called an "artifact," that is to say, of having created a situation in which what shows up at a rendezvous could not do anything else. Because in this case, it is she, as the organizer of the rendezvous, who would be responsible for what she thought she was simply exhibiting: the horror of discovering, for example, that one has been working with a solution that is "contaminated," into which one has oneself introduced what one then identified there! And therefore our experimenter might consider with perplexity an apparatus of experimental psychology in which the "subjects" are there in order to answer questions, in which consent is not a success but a condition for the experiment. She will ask herself if there is any rhyme or reason to an apparatus in which a response is guaranteed because those who respond have no choice, either because they can do nothing else (they are "trapped like rats" in a Skinner box) or because they agree, on principle, out of politeness or habit, to answer zealously all the questions it might please the experimenter to ask them.

The experimenter is perplexed, something that has nothing to do with the condescending observation about the distance between the psychologist's laboratory and the experimental ideal that defines physics, and which every science should try to obey. She is perplexed because for her it has nothing whatsoever to do with experimentation, because the "objectivity" that commands the laboratory of experimental psychology has nothing to do with the objectivity that her colleagues associate with the experimental success, which, should it happen, makes them dance. If scientists dance, they do not do so because they have been "objective," because they have measured in a reliable way, because their statistics are not biased. They do not dance to celebrate their own excellence but a success, which, while certainly requiring that humans activate themselves, think, imagine, discuss, does not depend just on humans.

If sociologists accepted the really very particular, highly selective character of what the force of experimentation presupposes, they might eventually not feel themselves professionally tied to "demystification." They might escape from the trap that is constituted by the alternative "us or them"—their "independent" or our "socially constructed" reality. Such an alternative supposes that the physicists and the sociologists are dealing with the same reality, in relation to which they are rivals. Now, sociologists of (experimental) science do not interrogate neutrinos, whether or not they have a mass, but very particular communities, such as those who were so excited at the possibility of concluding that "neutrinos have a mass." Their question bears, or ought to bear, on the manner of approaching the very particular regime of existence of such communities in a pertinent manner. These communities are certainly defined "socially," but the social must, in their case, include the active, practical, ambition of fabricating situations that authorize the enthusiastic conclusion, "We don't have any choice; we have to accept it; there is the fact!"

If one thing seems sure, it is indeed that such a community itself will never be able to satisfy the requirements of experimentation. If there is to be a successful rendezvous between experimenters and sociologists, this will not be a rendezvous of the experimental kind, which would allow sociologists to celebrate the fact that "the experimenters have spoken!"—that is to say, have been staged in a mode that "makes a difference" and puts the sociologists in agreement. The undertaking that seeks to resemble this kind of success—by "showing" that the practices of scientists respond, like every practice, to sociological categories—is as far away as is possible from a successful rendezvous. It's a judgment, and this judgment will be understood by those it concerns in the manner of a declaration of war.

It is possible that in this instance, the sociologists are victims of the traditional opposition between "fact" and "value," between what is supposed to impose itself on humans, whatever they think, and what arises from the manner in which humans, in every epoch, in every society, or in every

culture, make what they are dealing with matter. They wanted to show
that scientists cannot claim to stick just to the facts, which would be neutral in relation to values. They weren't wrong: values, in the sense that they
designate what matters, really are constitutive of experimental facts. But it
is not a matter of general values, which would "bring" scientists "back" into
the common realm of the mortals, all of whom alike are a matter for sociology. I have never encountered the "common realm of the mortals," but
what I call practical engagement signifies that the values of experimenters
as practitioners, without which their facts would not exist, are also what
make them diverge in relation to this common realm. Their obligations are
rather "uncommon."

But it must be said, and said out loud, that scientists themselves rely
on the traditional opposition between facts and values. When physicists
speak of an "independent reality," they generally borrow a vocabulary of a
philosophical, and not a practical, kind. They do not speak of "reality insofar as it makes *us* agree on this or that point" and often prefer to speak of
"reality insofar as it is also that of everyone else—even if it is we physicists
who have learned how to characterize it." This style—which is that of "scientific visions of the world"—is the manner proper to physicists, and in
the first place, to physicists celebrating their successes, of insulting everyone else, presenting reality such as it matters to them, such as it satisfies
their requirements and allows them to meet their obligations, as that of
everyone. Worse, as "explaining" what everyone deals with.

Here scientists reading me will prick up their ears. Is she going to prohibit us from leaving our laboratories, that is to say, is she going to link
us to facts that are certainly well-established but whose signification is
restricted to the context of the experimental rendezvous? Is she going
to deny that we have the capacity to explain processes that occur "in the
world"?

Explanation is not to be proscribed, of course. It can be conceptualized
as immanent to the success of a bringing into relationship. To create a
successful relationship with neutrinos, it is now necessary for detection apparatuses to take into account the consequences that physics has enabled
to be associated with "having a mass." The results of Davis's apparatus are
therefore explained, and they also concern the Sun as the locus for nuclear
reactions that produce neutrinos. But some scientists will demand more.
For example, would I deny that the behavior of a living thing can be explained in terms of "molecular interactions"? If I deny it, I have lost and
have been unmasked: my fine discourse constitutes nothing but a ruse,

worthy of a philosopher, to restrict the force of experimentation to what doesn't bother anyone, and not to questions that preoccupy everyone.

As I don't want to insult scientists, I will take a favorable case, one where it is not a question of submission or obeying instructions, which prevails whenever a scientist tries to ask a question that concerns everyone, and particularly those to whom it is asked. Let's take the behavior of an insect attracted by the odor of a flower. It is perfectly possible to test experimentally the hypothesis according to which a particular chemical molecule, one obtained by purification in the laboratory, attracts the insect in the same way as does the flower. But what does this success say? Evidently it doesn't give the slightest explanation of the insect's behavior: this is presupposed and now finds itself articulated no longer to the odor of a flower in a garden but to a molecule, the structure of which can be identified. We could certainly envisage mobilizing these insects, which buzz and fly around us, as "detectors" of this particular molecule, but the molecule doesn't explain its own detection.

Let's go further. Let us suppose that it is possible to identify the "receptor" that the molecule "activates" in the insect's brain. As in physics, sophisticated apparatuses, which have proved their reliability, will then intervene and enable the passage of the insect from the buzzing field to the laboratory. But that is not why the interaction between the active compound and the receptor will "explain" the behavior of buzzing insects. This behavior has instead been bracketed off, because in this new environment, the molecule is no longer what "attracts" but what is susceptible to interacting with an assemblage that is also molecular. The question has changed; it can allow the interest of a particular receptor to be identified, which will doubtless raise new questions. But the buzzing behavior is always presupposed: it is what "makes" both the molecule and the receptor "interesting." Certainly it may happen that the molecular/behavior link gets more complicated and that the buzzing behavior becomes much richer than the abstract definition with which "we" (that is to say, the specialists) had identified it. Here I have restricted myself to emphasizing the heterogeneity of questions that are linked together, but we should never forget the surprises that the brain, even if it is that of an insect, reserves for those who interrogate it. Thus the interesting receptor will doubtless not be the end of the matter; it will itself probably refer to a veritable labyrinth of cerebral couplings. And perhaps this labyrinth will inspire new questions among ethologists with regard to what insects are capable of. In short, where the scientific vision of the world affirms that molecular interactions "explain," we are dealing with a weaving between heterogeneous practices, an *art of consequences* punctuated not by "and therefores" but by "so then whats?"

To scientists who now oscillate between perplexity and indignation, I would say that it is their "vision" of a reality that can be explained in terms of atoms, molecules (or whatever you will) that is an insult to their own work, that dissimulates the very particular success that the fact of being able to say "this is explained by that" constitutes. Each time, what must be heard is that "this" has been rendered capable of testifying to the role of "that." Every vision of the world amalgamates and places what has been obtained through successful links, the selective establishing of relationships, the creative grasp of relays articulating heterogeneous assemblages, on the same homogeneous plane—the molecular plane, for example, from which the behavior of the insect might be deduced. The cost of the authority of the scientist who "explains," where others "believe," is a doubly mendacious language. With her colleagues she will assemble, hesitate, and even, as the case may be, dance. But when she addresses the "public," she speaks in the name of a reality that would, in itself, have the power to arm its representatives against illusion, a reality that would make of her a judge.

Not insulting experimenters is, for me, to understand fully what made David Wark rejoice, celebrating the fact that neutrinos henceforth had a mass: he and his colleagues had "learned something about the universe that nobody knew before." Something new, something that will provoke new questions, will add and not reduce. And it is this very particular joy that it is a matter of thinking, in order to resist the consensual reasons of Science. It is a matter of resisting at one and the same time those who identify Science with the conquest of a knowledge of reality that is objective at last and those who identify it with a conquering power against which it is a matter of struggling.

CHAPTER THREE

DISSOLVING AMALGAMS

The Enemies of Our Enemies

In a very fine article entitled "My Enemy's Enemy Is, Only Perhaps, My Friend,"[1] Hilary Rose, a pioneer in England of the feminist questioning of the sciences and involved in the Radical Science Movement,[2] spoke of her refusal to be mobilized by the "science wars" in their English version. That is to say, by a confrontation between, on the one hand, scientists "defending rationality," and on the other, proponents of what Rose calls the "C&P sociology of scientific knowledge," named after Harry Collins and Trevor Pinch, the two principal spokespersons of the critique of this rationality as a socially constructed myth.

Dismantling this myth, demystifying the authority of the sciences, however, would seem to correspond to a reprising of the politicization that feminists and "radical scientists" were aiming at. Collins and Pinch's book *The Golem: What You Should Know about Science* is intended for the "profane citizen" edified by epistemological vignettes that have taught him to nourish a respect for and confidence in science. Addressing oneself in this way to the citizen seems all the more praiseworthy in that the authors are careful to oppose the excesses that a systematic skepticism might lead to. "Claim too much for science and an unacceptable reaction is invited. Claim what can be delivered and scientific expertise will be valued or distrusted, utilized or ignored, not in an unstable way but just as with any other social institution."[3] The golem of legend, as Collins and Pinch present it, doubtless wears the word "truth" on its forehead, but what matters to the two sociologists is to remind us that the golem does only what it can. At best, it tries to do what is expected of it, but it is irremediably clumsy. It must be monitored. It's the same with the scientific golem, except that the myth of a "clean and correct" science allows this golem to avoid surveillance, allows us to forget that "a golem, powerful though it is, is the creature of our art and of our craft."[4]

"Science is not to be blamed for its mistakes; they are our mistakes."[5] Pinch and Collins can thus conclude. One might be tempted to adhere to this democratic wisdom, but one would thereby be adhering to a singularly apolitical point of view. "Our" art? "Our" errors? I've also happened to write "we," but that was to designate those people, of whom I am one, who belong to a history and a tradition (we moderns) the self-evident status of which needs shifting, whom it's a matter of making feel that they do not have the anonymous character, free at last of all credulity, to which they lay claim. The "we" of Collins and Pinch's "our" art and "our" know-how is, for its part, indeed anonymous. It is there to affirm first and foremost that the scientist is "a human, just like everyone else." As Rose notes, "At one level I cannot see why C&P and the mainstream [sociology of scientific knowledge] in Britain are so under attack, for they never question, as feminists and radicals do, the larger political role of scientific knowledge such as sociobiology, or, for that matter, the new genetics."[6] Nor are they interested in the "privatization" of research, because their primary target is the illusory autonomy of all research. "We" must monitor the golem, but the targets of this monitoring are the claims of "scientific knowledge" in general. Whether it is privatized or public matters little, as all knowledge is equally "socially constructed."

For Rose, the amalgam that makes it possible to speak of "Science" whether or not it is a golem, has a painful memory. The movement contesting IQ testing, which she participated in, was torn apart and crushed by the opposition between scientists who considered that it was "bad science" and those for whom, already, all science being a social construction, it was simply a matter of a science "like any other," the success of which depended, like any other (science), on a social relation of force. The irony with which the "social constructivist" enemies of IQ called on the "scientific" enemies of IQ to define what a "good science" would be, is, for Rose, that of professionals who were as authoritarian and sure of their right to do so as the scientists specializing in IQ whom they criticized.[7]

This is the kind of disaster that threatens the feminist movement if it doesn't learn to address scientists and, above all, feminists who work in the sciences, without taking them hostage, without calling on them to make a choice between their feminist and scientific commitments. "What separates the feminists from the mainstream professionals is that feminists are committed to building alliances, not least with other feminists, and consequently are very sensitive to the delicacy of the relationship between the feminist critics of science and feminists in science."[8]

One could call "positivist" any position that addresses "science" or "scientific knowledge" as if it were endowed with an identity, whether that is to praise it, to accuse it, or to make it the object of . . . science. By contrast, I would call Rose's position with regard to the sciences, which is also mine, "pragmatic": betting on the possible and not on the security of denunciation. Evidently pragmatism doesn't signify that "what one thinks doesn't matter, what counts is not to scandalize those who could become allies." If pragmatism, in William James's sense, is an "art of consequences," this is, as I have already signaled, in a demanding sense: to pose the question of the consequences of an idea, of what it exposes one to, what it commits one to, is not to "reduce" an idea to its consequences. It is to refuse the abstraction that claims to separate an idea from its consequences, that attributes to it a "truth" transcending its consequences. If scandal is a foreseeable consequence of an idea, that makes it as such part of the idea, part of the verification that makes the truth of this idea.

In this instance, the "positivist" position with regard to the sciences confirms, from a pragmatic point of view, the power of the amalgam that makes it possible to refer to "science" in the singular (or to "technoscience"). Because this reference is effective in both cases, whether one praises or denounces "science." It unites Rose's enemies and the enemies of her enemies.

To succeed in escaping the position that the amalgam assigns its critics—that of reinforcing what is criticized—doesn't signify defending a "pure" science unjustly amalgamated with false or impure sciences. It signifies the adoption of a position that has as a wished-for consequence that scientists caught up in the amalgam, notably feminist scientists, can experiment with the means of weakening it (the amalgam). That is what I am trying to do by betting on a possibility that requires the dissolution of the authority of "science" but certainly not the destruction of practices that succeed in conferring on facts the power to make a difference. The positivist-critical approach of the sociologists can have as many ready-made facts in its favor as one likes: I mean to resist it, because it is destined to bind scientists together and thus to crush a precarious possibility, a possibility that requires that scientists themselves separate from their reference to "science."

After the "linguistic turn," in America today one speaks of the "practice turn."[9] *Practice theorists* are united by the concern to resist the exclusive privilege of propositional knowledge ("knowing that"). They emphasize the importance of competence and skills ("knowing how") that make for the specificity of each practice.[10] For the theorists of the practice turn, it is thus notably a matter of dropping the polemical relationship organized around epistemological vignettes articulating ready-made facts and theories. Practice refers to sciences "in the making"; it encompasses the perfecting of instruments, the writing of articles, the relations of each practitioner with her colleagues, but also with everything that, and everyone who, does or could count in her landscape. Nothing is "ready-made." Everything is to be negotiated, adjusted, aligned, and the term *practice* designates the manner in which these negotiations, adjustments, and alignments constrain and specify individual activities without, for all that, determining them.

Practice theorists thus want to escape as much from the fiction of a "knowing subject" as from that of a structure or a field that would identify the roles that actors come to play. A practitioner of the sciences doesn't produce a knowledge answering to a canonical objectivity that would define the "subjects" of this knowledge as interchangeable; she wonders if her colleagues, whom she knows well, will agree to take into account what she is proposing. In so doing, she doesn't play a role she is assigned either; she wonders what role she can risk. In short, she becomes active, in a "field," for sure, but one that doesn't dictate her conduct to her. Rather, it is a field of constraints, which doesn't so much say what must be done but what can be risked, the "moves" that can be envisaged.

Having said that, if there is indeed a "turn," the direction that is to be abandoned is much more clearly defined than the direction to take. Can a practice be reduced to individual habits and know-how? The practitioner would then "find her bearings in her landscape" a little like every one of us who finds our bearings in a specific social situation: one knows what one can do or say, how far one can go, the gaffes that must be avoided, the necessary life skills. Or must a practice be explicitly described as referring to a collective and to the negotiations that hold it together? While the majority of life skills, in referring to what they exclude, make no explicit reference to what they make possible—"you just don't do that; that's all there is to it"—do not scientists have an explicit concern for the collective to which they belong and for what makes it hold together? And in this case, doesn't negotiation make norms, which would be imposed as such on members of this collective, intervene? In other words, can an infraction of such rules be assimilated to a "gaffe," or will it be judged and condemned as putting the practice in danger?

The question of the commitment of the theorists of practices themselves is added to these "theoretical" questions. Should they, as is appropriate whenever one is dealing with natural phenomena, for example, adopt a position of exteriority in relation to what they are studying? Should each describe the manner in which "their" practitioners busy about, fabricate, recruit, argue, a bit like the manner in which a specialist describes an ant's nest?[11] Or should they emphasize that it is "our" world that these "ants" contribute to constructing, that it is the manner in which we live, think, understand things, and understand ourselves that is at stake? But in this case, wouldn't those who are interested in scientific practices have to abandon a position of "epistemic sovereignty," which gives itself the right to define scientific practices as "their object"?[12] What they study is then effectively a *matter of concern* in Bruno Latour's sense, that is to say, a collective and political matter.

To protect oneself from a so-called positivist position, it is not enough to undo the amalgam that identifies "science" and "knowledge" and to make way for "practices"; one must also envisage the risks of what would be a skill at "undoing amalgams." In this instance, the "positive" sociologist who follows scientists who busy about in the manner of ants and the committed thinker who is concerned with what sciences "do" to the world seem to me to be exposed to two distinct risks. On the one hand, following the manner in which scientists and their allies not only construct claims about the world but also the world itself, our world, can absorb attention in such a way that the sociologist comes to appreciate their undertaking in terms of the scientists' categories themselves: woe betide those who suffer, who are reduced to silence, or who never had a voice to make heard, because the only people to count are "the protagonists who made themselves matter," who had the power to object, who had to be recruited, whose interests and reasons one has to recognize and accept.[13] But on the other hand, the commitment implying that scientists are not ants, that the power associated with science must be called into question, exposes the one who endorses it to the risk of being transformed into a righter of wrongs, into a denouncer of the imperialist vocation of science, into a spokesperson of what is driven into the shadows of nonscience or nonrationality. In both cases one falls back into a "positivist" position: the sciences, such as they are made, find themselves attributed an identity that justifies the interpreter's position.

And the decision to adopt a normative approach to scientific practices is not sufficient to resolve the problem. Certainly, the norm signifies that scientists are committed, that they know themselves to be committed. It can also allow the posing of the problem of the norms to which those who describe are themselves committed. Finally, it can allow the declaration "we are not ants!" to be heard. However, the defect of the notion of the norm is to rhyme a little too directly with that of "submission." A norm is imposed, one is bound by it, one vindicates it, one identifies with it. Certainly there are scientists who vindicate a normative approach, a "one must" that justifies what they take into account and what they exclude, and who make use of these norms to police their ranks, to chase off as "nonscientific" any proposition that doesn't refer, for example, to a manipulable and measurable object. But to take it as "normal," to assign a normative identity to "science" fatally ends up in a staging of conflicts between opposed normative identities with, possibly, the necessity of arbitration in the name of some transcendent value. This position is occupied today by the omnipresence of "ethics committees," now the ritual arena for "conflicting values."

For the "committed" sociologist, the pragmatic price to pay is high: if one considers scientists as a priori subjected to norms, how is one to try to address them with a bet on possible alliances, with a bet on the possibility of modifying "science such as it is done"? How is one to avoid taking them hostage in a "conflict over norms"? This is the price that Hilary Rose refused to pay. However, the problem changes if one admits that the question of the norms to which practitioners might eventually have to submit doesn't constitute a question for the theorist of practices, who would have to detect and identify them. It is a question posed by and to *the practitioners themselves.* It's a question about which they hesitate, become divided, and oppose each other, as can happen. "Do we have the right?" is a question that intervenes in the controversies and conflicts between scientists. Attributing norms to scientific practices amounts to "knowing better" than those who hesitate, become divided, and oppose each other.

From the point of view that I am proposing here, scientists are therefore not ants, at least, not in the sense in which they are defined by the ethologist who, by careful manipulations, ruses, and artifices, endeavors to make explicit what it is that their behavior obeys, including what makes them "hesitate." The hesitations of scientists, their choices, the range of arguments covered in their controversies do not constitute the manifestations of the "normative reasons" to which they would subject themselves as scientists either. On the contrary, it is the scientists themselves who seek to make explicit what it is that makes them scientists.

Thus contemporary specialists in superstrings defend the scientifically legitimate, even necessary, character of their undertaking, in spite of its eminent obscurity, by affirming that physics requires a tenacious faith in the intelligibility of the world. If one gives up on superstrings, they argue, one also gives up on making the diversity of physical interactions intelligible, signifying the loss of the faith that makes physicists think and create. But the identity that is thus accorded to physics forms part of a register of argumentation destined to impose one conception of physics against others: the link between physics and faith appeared at the start of the twentieth century under the pen of Max Planck in order to disqualify another physicist, Ernst Mach, for whom physics was linked to critical rationality and should thus rid itself of models that entailed a reference to unobservable atoms of a metaphysical origin.[14] The necessity of a "faith" in the intelligibility of the world thus translates the capture of a historical analysis—the "great founding figures" (or some of them) testify to such a faith—and its transformation into a normative argument internal to physics.

Even the "deconstructionist" thesis according to which that which scientists call "discovery" in fact constitutes an "invention" (a human fabrication) has already been captured and used in an entirely "constructivist" manner in the argumentation that has allowed for genes to be patented: a decoded gene sequence can be patented as an "invention" because it is inseparable from human procedures and techniques. In a more general manner, the entire development of biotechnology has been accompanied by a grand discourse celebrating a science freed from the questions of faith and "metaphysical" intelligibility that encumber physics. Laboratory biology would replace physics at the summit of the hierarchy of the sciences and would propose new "antimetaphysical" norms, against the idea of a world "discovered" by the sciences.

Whatever the normative definitions or anthill descriptions produced by practice theorists may be, they are exposed to captures of this sort, and this process of capture is what gives the "positivist trap" its power. Scientists effectively construct the identity of the sciences incessantly, that is to say, they borrow from and propose to those who describe them the finally adequate categories that they obey. From this point of view, one could treat them as anti-chameleons: more or less any identity can, circumstances permitting, be "recuperated," but on the condition that it allows them to remain distinct from the landscape, to impose the perception of a difference that must be recognized.

From this point of view, the pragmatics of alliances defended by Hilary Rose are very pertinent: to address oneself to scientists in a mode that deliberately, constructively, avoids taking them hostage in a conflict of values is to know that the values on display are, in the first place, rallying signs destined to mobilize, to fabricate a collective identity that will allow all of those who would pose questions to be denounced as "traitors."

However, it will not be concluded from this that the sciences are "simply a construct." The list of what can be grist to the mill may be endless, but that doesn't signify that the mill itself is just anything whatever, or that the manner in which it grinds flour is arbitrary. That is why I have proposed that a practice be characterized in terms of two aspects that the notion of the norm does not allow to be distinguished: *requirement* and *obligation*. The identity that a science affirms is always affirmed in the framework of an offensive or a defensive operation; it makes explicit requirements that it is a matter of having recognized; it silences critics or rejects challenges. But the question of their obligations does not mobilize practitioners; it is neither offensive nor defensive. It makes them hesitate.

The distinction between requirement and obligation and the correlative proposition of approaching norms as amalgams that can be undone are certainly not based on what Latour calls *matters of fact*, on facts that would give them an authority. They do not have for their stakes the best manner of "correctly" describing practitioners but the manner of addressing them. And, what is more, of doing so under the sign of an unhappily probable future in which the idea of "colleagues" with whom it is appropriate to hesitate could well exist only in the disenchanted memory of times past. They thus respond to what Latour calls a *matter of concern*, my concern here being the possibility of addressing a practitioner without identifying her with what she requires but insofar as she is "obliged" by her practice. My concern is the possibility of betting that she is able not to take shelter behind an identity.

In effect, an obligation does not identify, because it leaves open the question of knowing how it must be fulfilled or of what would betray it. It does not have as its vocation the gathering around the same mode of judgment, and it can divide. Moreover, the question of what obliges practitioners is never general. It is always a matter of what such and such a situation or proposition obliges them to do. But it also demands that those who attempt to approach practices not "know," when practitioners themselves hesitate.

Such, then, is the meaning that I will, for my part, give to what is called the "practice turn." This turn will have to break with any ambition to neutrality and scientificity or "epistemic sovereignty." It demands that those who are interested in the sciences accept to practice the pragmatic art of thinking with those consequences the verification of which ought to be called creation: a creation of new kinds of links with scientists. That is to say, that they practice the pragmatic art of conferring on the possible as such the power to oblige thinking.

"Realist" Obligations?

The pragmatic art of consequences has nothing to do with the art of putting together a proposition in such a way that it results in the desired consequence. Nonneutrality doesn't signify the arbitrariness of a will that bends what it is dealing with as a function of the end it pursues. This is something that, in any case, experimenters well know when they say of a hypothesis that it is "ad hoc," proposed so as to resolve a difficulty, to bring a troubling result back into order, and having no other consequence than that. By contrast, a hypothesis that is "interesting" ought to allow other consequences than those that designated the direct stakes of its formulation to be envisaged.

The consequences attached to the notion of "requirement" will be envisaged in what follows, even though some of its stakes are already clear: it will be a matter of giving them their full importance without diminishing the difference that still exists today between the "it works" that might make one dance for joy and the "it works" that will satisfy a sponsor or will authorize a patent. And it will be a matter of doing so in a way that incites scientists to think and to feel that there are other manners of resisting than those that situate them in a besieged fortress, faced with a world that is generally irrational and therefore hostile. I have chosen to deal first with the question of what "obligates" scientists as scientists because an amalgam threatens. So far I have, in effect, only spoken of experimental, or theoretico-experimental, practices, and I am, as a consequence, in danger of reproducing the habitual split between "true" sciences, those whose success might be said to "confer on what one is addressing the power to decide between its interpreters" and all the rest, who would be obligated by nothing.

Already, when it is a matter of the field of theoretico-experimental sciences, this formulation of scientific success doesn't correspond to something empirically evident but to an engaged reading. It effectively gives a strong meaning to what can be called "experimental realism," the possibility, typically, of affirming that what they have entered into relationship with "really exists." Now, these fields have not been lacking practitioners who invoked a more "rational" way of operating, attentive to separating scientific descriptions from any "ontological" scope. One could even put a "date" on experimental realism, which only became dominant in the twentieth century, when physics and chemistry succeeded in demonstrating that they were capable of "going beyond phenomena" and populated the world with the "experimental beings" that are atoms, molecules, photons and, subsequently, neutrinos, and now the Higgs boson. With the notion of a "guarantor," have I not established a link between experimentation and realism that prolongs the amnesia of today's scientists with regard to the determined antirealism of certain of their predecessors?

To be sure, it would be easy to answer that from the point of view of contemporary scientists; such predecessors have been "defeated" precisely because of their antirealism. To reactivate their memory would then be seen as a polemical act, translating my intention to demoralize the inheritors of those who, at the start of the twentieth century, could affirm that they were henceforth capable of "going beyond phenomena." But there is another reason for bracketing off this possibility of maintaining a position of "skeptical lucidity" in relationship to the theoretico-experimental sciences. Lucidity of this kind is not the inevitable guarantee of a position that opens up the question of other practices, answering to other obligations. This is what I would like to show with regard to the "conventionalist" position of the mathematical physicist Henri Poincaré.

It was with regard to what is doubtless the most brilliant success of nineteenth-century physical chemistry—the affirmation of the conservation of energy—that Poincaré defended this position. From the outset, the conservation of energy has been synonymous with the realism of experimental physics. Max Planck wrote that if other civilizations, elsewhere in the universe, have developed a science, they too must have discovered that the entirety of natural processes are united by the fact that they conserve what we call "energy." Now, Poincaré dared to maintain that this conservation is a convention: energy is a "something" that physicists define as constant, but they cannot demonstrate this. Indeed, physicists postulate this conservation, and it is on that basis that they define what energy is in each case concerned. More precisely, Poincaré showed that when the quantity of energy associated with a process is not conserved, physicists do not conclude that conservation is refuted. They make other, as yet unidentified, forms of energy intervene and restore conservation.

Poincaré, though, was lucid and not "deconstructive." There was nothing arbitrary about the convention for him, because maintaining it depended on its experimental fecundity: hitherto, every time that the conservation of energy seemed to be refuted, hypotheses that implied a new form of energy that hadn't yet been taken into account were crowned with success; on the day when such hypotheses would be merely ad hoc, having no other consequence than energy conservation, the convention would have to be abandoned.

When French Christian thinkers—who might be considered the ancestors of contemporary deconstructionists—proclaimed the bankruptcy of science at the start of the twentieth century, they used Poincaré and assimilated his convention to a purely human, artificial, and contingent creation. Poincaré was appalled, and, at the end of *The Value of Science*—an introduction devoted entirely to denying any kind of link between his theses and relativist skepticism—he affirmed that what we have access to via the harmony expressed by the laws of mathematical physics is not a reality that is completely independent of the mind that conceives it, for sure, but is indeed "objective reality," in the sense of that which could be common to all thinking beings. According to Poincaré, it was the universal harmony of the world, the source of all beauty, which the slow and painful progress of science gradually makes better known to us.[15] A medieval theologian accepting the thesis of God the mathematician could not have put it better, and in one step, we have passed from apparent skepticism to the celebration of physics as transcending all other knowledge by virtue of its truly symbiotic relationship with mathematics. Put on the defensive, the specialist in mathematical physics has become the aesthetic brains of humanity!

This example translates the instability of skeptical lucidity, especially when it stems from a mathematical physics. Whenever agnosticism or nonrealism make mathematics prevail in its own right, they are always susceptible to calling on the progress of physics as a witness and brutally resuscitate general claims that make this science the privileged interpreter of the world. And these claims give arms to claimants. Thus certain contemporary cognitivists, after a closely argued deconstruction of the realist claims of science, suddenly propose that perhaps biological evolution has configured our brains in such a way that there exists a kind of "affinity" between their mode of functioning and that of nature: from the moment that the sciences are justified by "the functioning of our brains"—the object of study of cognitivism—and not by the arguments constructed by scientists, honor is saved. Curiously, here, too, it should be emphasized, it is the success of physics alone that justifies the affinity hypothesis. What is more, it's a matter of physics as producer of "objective visions," of representations with a mathematical appearance, abstracted out from successful experimentation.

Basing the obligations of physics on the success of experimentation, as I do, is to note that critical, and therefore antirealist, readings of knowledge regularly end up conferring a veritably hyperbolic authority on physics' representations of the world. Although successful experimentation is an event, representation appears as following from a right that arms judges who are indifferent to the "trivial" detail constituted by the difference between a question posed unilaterally and a "good question." This difference does not deny the importance of mathematics but rather the idea that the success of physics derives from a "mathematization of the world"[16] that it would be enough to prolong.

In fact, my choice of linking the success of the theoretico-experimental sciences to the possibility of designating what they interrogate as "responding," having obtained the power of confirming an interpretation against others, corresponds to the manner in which the hesitation between realism and antirealism in the history of the physical and chemical sciences was settled. It was by a deliberate use of the "force of experimentation" that Poincaré's interpretation and that of other antirealists of the time was consigned to epistemology, that is to say, to what may interest philosophers but cannot make scientists hesitate. It was effectively on "atoms" and other unobservable particles—defined by critical physicists and chemists of the nineteenth century as, at best, auxiliary aids to the imagination— that the power of silencing critical skepticism, of putting an end to any possibility of antirealism, was conferred.

In 1913, the publication of Jean Perrin's *Atoms* was the victory song of a science in which one dances in laboratories. Perrin not only succeeded in showing that one could "count atoms"; he also succeeded in making the nature that he interrogated the guarantor of experimental realism: science is not an ordering, however harmonious, of the regularity of phenomena; it actually goes "beyond phenomena"!

Perrin certainly "danced" when the figures resulting from thirteen distinct experimental situations took up their places in thirteen equations, each one of them modeling the corresponding situation in a manner that made explicit the hypothetical role played by "unobservable particles." He was effectively able to obtain "practically the same" value for "Avogadro's number" thirteen times. "Our wonder is aroused at the very remarkable agreement found between values derived from the consideration of such widely different phenomena" he wrote.[17] But for Perrin, this agreement evidently had nothing to do with a miraculous coincidence. Nature really did play the role of *guarantor*, corroborating the pertinence of "atomic" descriptions. Nature came to the rendezvous that had been prepared so as to confer on it the power to make the difference that would authorize Perrin to silence the critics.

But the example of Perrin, filled with wonder, is also what can protect us from the temptation to amalgamate and so come to some general definition of "Science" (and the resulting hierarchy, with experimental sciences being the accomplished model, the others an approximation). Because the type of success that made Perrin dance cannot be generalized. Rather, the "miracle of agreement," the "it cannot be by chance" that is able to generate faith in the experimental laboratory designates instead what may be a trap for other scientific practices, those which, in this case, work on a terrain, on "indices."[18] In these fields, the "it cannot be by chance" may become the beginning of what practitioners acknowledge as the greatest danger, the moment—nicely narrated by Umberto Eco in *Foucault's Pendulum*—when nothing is insignificant any longer: the terrain on which one seeks to reconstitute "what happened" on the basis of traces and indices becomes extraordinarily prolix, as if what was asked to play the role of guarantor came into play and lured the enthralled researcher into a delirious spiral.

What makes practitioners who work with indices hesitate doesn't have a great deal to do with what makes experimenters hesitate. The obligations diverge, one's success being precisely what the other's must be wary of. In any case, this divergence seems well-enough established for me not to dwell on it, except to emphasize that it is often translated in such a way that specialists get accused of dogmatism, because they don't know how to "say clearly" why they are wary of the enigmatic indices that fascinate amateurs.

In Egyptology, for example, the pyramids are reputed to be "evil objects," likely to make those who seek to solve their mystery go mad, but the wall of silence that Egyptologists establish in opposition to indexical constructions is a quick-fix solution that reinforces the general image of a science struggling against irrational opinion.

Entirely different again is the situation of the fields of research, from animal and human psychology to sociology and pedagogy, which the psychoanalyst Georges Devereux, originator of ethnopsychiatry, grouped together under the name "sciences of behavior." What Devereux, in effect, called "behavior" was whatever is capable of "addressing itself" to a milieu and of conferring a signification on it. Here the question is no longer, "What makes scientists hesitate?" but rather, "Why don't they hesitate?"

What Are Those We Interrogate Capable Of?

One doesn't dance in the laboratories in which those whom Georges Devereux characterized as specialists of human or animal "behavior" attempt to prolong the experimental model. There, one applies a method that is supposed to guarantee objectivity and would therefore be scientific. But far from generalizing the experimental undertaking, this method betrays its obligations. It is no longer a question here of creating a relationship but rather of ignoring, destroying, or hijacking the relationship that the "experimental subject" already entertains with his world, and of acting as if the subject was, like an atom or a neutrino, indifferent to that operation.

The reference to obligations here will change its meaning. It will no longer guide a manner of "saying clearly" what makes practitioners think and hesitate but will entail a calling into question of certain activities that present themselves as scientific practices. There's an obvious objection. This reference thus has normative consequences! It doesn't have as its consequence a manner of addressing oneself to practitioners but rather it contests the way they identify science with a respect for a method.

How to address those who present themselves in the name of a method? There are many undertakings that are based on a method that it matters to respect. An accountant has a method, and schoolchildren are supposed to learn to have one. The value of a method is not as such contestable. One could even say that every method translates the organizing capacity of language when it assembles a situation as a function of an end—first do this, don't forget that . . . But what is the end pursued by scientists who make scientificity and respect for a method coincide? Would not this end be the label of science itself? And isn't the cost of this the a priori exclusion, in the name of method, of what might entail a difference between experimental facts, those that allow one to "learn something that nobody knew before," and "methodical" facts? This signifies that hesitation is an enemy and that all who take an interest in the manner in which obligations make scientists hesitate will, similarly, be enemies. The battle lines are set.

In truth it's impossible to address scientists "of method" on the basis of their obligations; one can only raise the problem this kind of a science poses. This is where Devereux's approach is pertinent, because it addresses the question of what method "does" to those who methodically study behavior and identify such studies with an experimental undertaking. How does this use of method affect the way they think and imagine? A "genuine experimenter," obligated by success, by the creation of a relationship that allows for learning, would restrict himself to denouncing any resemblance with his own undertaking, to reminding us that if what is studied is a behavior that confers a signification on its milieu, then this behavior will never be able to assume the role of "responding" to a question, confirming the pertinence of this question. Because what it will respond to will be the question itself or more precisely the signification that the question confers on the milieu it organizes. For his part, Devereux does not denounce what is produced in laboratories of behavior, in the name of the obligations of experimental proof. Rather than constituting the ideal of the proof as an unsurpassable horizon, he is interested in what singularizes the behavioral scientist who attempts to produce methodically correctly established, supposedly objective "facts," apt to prove. And for him, this singularity is an *anxiety* that neither the physicist nor the chemist experiences, the anxiety that can be provoked by the knowledge, which is insistent, even if it is denied, that for the other who is interrogated, the proof that is sought is an ordeal.

The anxiety that interests Devereux is marked not only by the manner in which one does not dance in laboratories of behavior but also by the manner in which questions that bear on what those beings interrogated "can become capable of" are excluded or disqualified. If "genuine" experimenters presented themselves well, any resemblance with method-based sciences would disappear, because these experimenters never stop enriching their beings with new "capacities," multiplying what it is that neutrinos, atoms, molecules, cells, bacteria, or neurons are capable of. The manner in which specialists of behavior "belittle" those they interrogate amounts, on the contrary, to defining certain capacities in a mode that disqualifies them. This is translated in two, distinct, manners. Either the capacity that an animal has of conferring a meaning on what happens to it is denied, and conceptions that reduce animal experience to a minimum are favored, alone defined as scientific, in the name of a struggle against squeamishness that the genuinely scientific attitude imposes. Or—particularly when it is a matter of humans to whom one cannot reasonably refuse the capacity of interpreting the questions they are asked—it is this capacity itself that will be methodically "belittled," defined as an obstacle to be overcome. The subject invited to the laboratory will thus often be "duped," misled about the question the experimental situation is about, in such a way that his response may be purified of what makes of him a thinking and interpreting being.

The history of experimental hypnosis has given me the opportunity to understand Devereux's cry, "What a worthy science of behavior wants is not a rat deprived of its cortex (figuratively or literally speaking) but a scientist who has had his returned to him."[19] Because this history has the particularity of staging a lengthy hesitation that bears on the possibility of reducing the "facts" regarding hypnosis to artifacts. The subject is not "hypnotized"; it's not the facts about "hypnosis" that emerge in the laboratory but facts relative to its cultural image and the manner in which the subject has interpreted and endorsed the role implicitly accorded to him by the experimental protocol.[20] A "worthy science of behavior" would thus have sought to approximate in a pertinent manner the regime of experience that is uncontroversially characterized, among other things, by the singular capacity of a hypnotized subject to integrate into his behavior the expectation that he perceives. But what has haunted researchers instead was the threat of being interested in a chimera: if there aren't any "facts that prove" the "state of hypnosis," one has to conclude that "hypnosis doesn't exist," that it's "nothing but role play." In fact, the experimental history of hypnosis has called into question in an exemplary manner the question of knowing whether the subjects of experimental psychology, rather than really being "duped," endorse the role of dupe that they are assigned. I do not wish to denounce the concerned scientists but to address them as victims, subjected to a model of science that has stopped them from thinking.

The question of what might oblige the "sciences of behavior" will remain open for now. What matters here is that such obligations can but translate an active divergence marking scientific practices. The practice of a science of indices allowed for a mode of hesitation that resembles the mode of hesitation that brings together the "laborers of proof," because the hesitation concerned the question, "Does this accumulation of indices 'really' allow us to affirm that . . ." In contrast, the possibility of speaking of practice with regard to sciences of behavior would call for this kind of resemblance to be abandoned, for the amalgam that identifies "science" with a production of knowledge *by* the scientist *about* what he interrogates to be dissolved. The creation of a relationship with a being for whom this relationship is not indifferent cannot be assimilated to the creation of an access that makes it possible to "obtain knowledge." If obligations can be formulated here, they will bear instead on what the relationship that is created will render *both its terms* capable of. Are they both mutilated, as Devereux emphasized against the sciences of behavior, or are they both, albeit in different modes, capable of making what matters for them count?

A question then poses itself: If they diverge to such an extent, can one still talk of "scientific" practices? What does the adjective correspond to if the substantive is refused? How is one to "speak well" of what brings these practices together through their divergences?

Perhaps we might hear David Wark again here, one last time, speaking of his joy at having learned "something about the universe that nobody knew before, and now you can say it!" Perhaps the meaning of the common term *scientific* is simply here, in this "learning something that wasn't known," something new, about what Wark calls the universe. In this case, the divergent character of scientific practices would translate what I will call a "generic" question: How can one learn something new? Learn and not produce, because the production of novelty is not characteristic of the sciences. Learn, in the sense that one has learned about and *thanks to* what one has addressed. The divergent plurality of the sciences, in the practical mode of manners of hesitating, would thus make the plurality of obligations and risks associated with this challenge exist. Learning something new in a mode that affirms, qua *matter of concern*, the importance of having to "address oneself well" to what it is a matter of learning from, of having to create a relationship with it that allows it to testify to what it is or what it could become capable of. Antimethod.

Let us repeat once more that the hypothesis guiding me is not neutral. There are many perfectly respectable practices—those preoccupied with collection, conservation, and classification, for example—in which the "new" doesn't have much meaning. In fact, in the acceptation that I am proposing, the adjective *scientific* is oriented in terms of the event that the "invention of modern sciences" constituted and not the meaning that the world *scientia* might have had previously, nor the set of knowledges produced by all the peoples of the earth, which are themselves collected and conserved today. "Scientific practices," such as I am proposing to approach them, are thus not privileged, but they are brought together as a function of a question that they themselves make matter.

On the one hand, this question is not general and has no general response. Any response that takes itself to be general will result in a hierarchization of the sciences. Thus, it is said that the physicist Ernest Rutherford once remarked that he knew only two types of science: physics and stamp collecting. The opposition marks well the crucial character of the word *new*: stamps are what preexist the collection; they are, for example, the "regularities" that it is sufficient to observe. For Rutherford, even chemists limited themselves to classifying elements, compounds, reactions. A physicist (Rutherford himself, in this instance) had to intervene in order that the chemical elements arranged in Dmitri Mendeleev's periodic table could be substituted with atoms capable of disintegrating (of jumping from one position to another in the table). At last, something new!

On the other hand, the question is generic in the sense that, if it is pertinent, any "scientific" practice should be able to present itself in a singular mode, translating, far from any general model, the singularity of what it addresses itself to.

Contrary to the adjective *general*, the signification that I am conferring on the adjective *generic* doesn't aim to describe or to judge. To propose a generic question is a pragmatic operation, because it demands to be verified by its consequences. In this instance, if it is fruitful, the question of "how to learning something new" should have as its consequence a double operation of the dissolving of amalgams. On the one hand, this question does not bring together "the sciences" without at the same time making them diverge; on the other hand, if practices that don't make this question matter are excluded, they aren't thereby judged. The adjective *nonscientific* is no longer an opprobrium. It is normal that what matters not always be "learning something new."[21]

The generic question "how to learn something new" thus effects a divorce of the sciences from their consensual reasons, which identified them with a general rational approach, opposed to the irrationality of opinion. It responds to what could be called a "chemical" pragmatics, in the sense that chemistry was, for a long time, the art of fire and of solvents, which function, like fire, to undo amalgams and aggregates in order to make possible an activity that they prevented.

To disaggregate, to dissolve, to insist on divergences: these could be especially important for the question that preoccupies the specialists in what is called, in the United States, *cultural, gender,* or *postcolonial* studies, the claim of the sciences to produce knowledges that matter for everyone and that, as such, are entitled to destroy, or to dismiss, through the category of "cultural belief," the ways of doing and knowing of other peoples and other epochs. The question of alliances about which Hilary Rose writes could be posed with these specialists if the generic characterization that I am proposing allowed for the questioning of this claim to be decoupled from the questioning of practices as such.

Where Science Leaves the Stage

I will take as my example here contemporary medicine, sometimes called "scientific" or evidence-based medicine. It's an interesting example because this medicine seems to be able to make the category of the "new" and the claim to interest everyone converge, under the sign of science. New medical treatments are constantly being put on the market, the result of lengthy laboratory work, and the question of "how to heal" matters everywhere, in every human culture. However, there is a polemical dimension to the enterprise. Medicine "armed with facts that prove" forces other therapeutic practices up against the wall: Bring your data! Show us your facts! One might be tempted to understand this polemic as a conflict of norms and of values, a temptation all the more powerful, given that the operation of forcing up against the "wall of proof" often induces modern, so-called alternative therapists to take a counterposition that imprisons them in conflictual order-words: spiritualists versus a so-called materialist science, holists versus a so-called reductionist science, and so on.

Let's start then from this "up against the wall" of proof, which makes the difference that should interest everyone between a "proven" medical treatment and an "alternative" remedy. What role do the sciences play in this differentiation?

Along the route to the creation of molecules that are candidates for the status of "new medical treatment," there is evidently much research for which the stake is to learn something new. But what matters here is that it really isn't this research that forms the basis of "up against the wall" polemics. The instrument through which therapies go up against the wall is nothing other than the challenge of having to succeed in the "clinical trial" on which the duly regulated passage from promising molecule to "new medical treatment" depends. And the success of the clinical trial doesn't matter to researchers alone but to all the protagonists: doctors, industrialists, financial speculators, as well as the patients concerned. It is therefore the clinical trial that should be the object of a "caustic" description, one able to disaggregate the generalities like "rational," "scientific," "objective" that make it possible to confuse scientific practices and enterprises that produce new entities supposed "to interest everyone."[22]

Caustic does not signify denunciatory. The clinical trial responds to a demanding method—one might even say it is testing (éprouvante)[23]—and to speak of a "trial" is not false, as the molecule stands as a suspect, as we will see. Furthermore, the legitimacy of such a method is perfectly defendable: it translates the fact that pharmacological tests, which differentiate between a remedy and a poison, are not enough, that public money, or that of private insurance, must be well-spent, not wasted on ineffective medication. By contrast, what doesn't have the slightest legitimacy is the link that is often established between a clinical trial and an experimental proof, that is to say, the idea that it is "scientific rationality" that allows for a decision to be made about a question that interests everyone. It really is a regulation that defines the "new medical treatment."

In a slightly caustic mode, one can say, more precisely, that however loaded with science the trajectory of a molecule with a pharmaceutical vocation may be, sciences leave the stage from the moment that the question of healing enters onto it. It is, in effect, at the moment of the clinical trial that "real patients" intervene in the trajectory of the molecule for the first time and the manner in which the scene is organized manifests, in an entirely explicit way, that what henceforth prevails no longer has anything to do with a laboratory situation, where biochemical mechanisms, cells, tissues, and so forth are being interrogated. What prevails is the risk that the patient gets better, is even healed "for the wrong reasons," that is to say, for reasons that have nothing to do with a specific activity that can be attributed to the molecule.

This is why the statistical method dominates in the clinical trial, with comparison of groups that have to be homogeneous and into which patients recruited for the test are distributed in a random manner. Some of them will absorb the molecule and the others, a "placebo," a product with no envisaged physiological effect. Such a method is perfectly indifferent to the questions that might matter to everyone with regard to the illness or to those that might matter to researchers with regard to the molecule's mode of action. It is incapable of determining if an observed improvement is "due" to the absorption of molecule, other than to consider that where the active molecule does come on stage, the other reasons for getting better (the wrong ones) disappear. In fact it doesn't constitute a *proof* but a *test*, in the sense of an ordeal, because what is being tracked down is a "lie": not that of "false cures" but that of a molecule, if the verdict is that it doesn't play any statistically detectable role in the process of curing or improvement.

The statistical method has to be blind, because its vocation is to filter, and it cannot, therefore, learn anything but can only detect a positive statistical correlation between the "taking of the molecule" and "improvement." But it is precisely here that another kind of lie often appears: the claim to the universality according to which "no matter who," whatever their culture or convictions, can benefit from the molecules that have been "tested/proved," just like, anywhere on earth, falling bodies testify for Galileo.

It is a matter of a lie because this "no matter who" has nothing to do with the Galilean "no matter where." This latter signifies at each and every point on earth, whereas the "no matter who" doesn't in the least bit signify "each and everyone" but rather a sick patient who exists nowhere because the definition is statistical. The correlation that is established effectively bears on groups of patients defined as homogeneous from the point of view of the illness targeted, all patients being treated in the same way, the only difference being that of "placebo or molecule," this latter being distributed double-blind, everything being suspected of "making a difference," the doctor who knows what he is administering included. "Universality" is therefore not a successful achievement but that which the method supposes and requires: short of bias or trickery—which can be denounced but which teaches nothing—the statistical verdict has to be blind to the fact that the patient is a man or a woman, American, Papuan, or Chinese, educated or "caught up in superstitions."

To "speak well" of clinical trials, one would have to speak of a "convention." To say convention is not to say arbitrary, in the sense of a "merely human decision." Clinical trials do effectively eliminate a good number of the molecules proposed; otherwise the pharmaceutical companies would not, as is sometimes the case, seek to cheat. But they do not produce the type of success that unites and obligates experimenters. Rather, it is a matter of getting protagonists with divergent interests—industrials, researchers, private insurance, or the state agencies who pay for the medicine, doctors, and nowadays also patient groups—to agree in a stable manner. The trial is what brings them together, that with regard to which they have "agreed." That is why, moreover, none of them is expected to be active: they must all be kept in suspense, awaiting the verdict of statistics, which they have accepted they will agree with.[24]

On the one hand, to affirm, faced with an experimenter, that experimental proofs amount to conventions, to agreements among colleagues, is to insult them. On the other hand, to affirm that clinical trials respond to a convention is to make matter the success constituted by such a convention, an agreement made between divergent interests, and the necessity of respect for the convention for all those who are party to it.[25] The success constituted by a convention must be carefully distinguished from any idea of rationality or normality or any other reference claiming the power to bring protagonists into agreement, because this success has to be protected and protecting it implies keeping active in memory that, in the case of a convention, no one has the power to bring the protagonists into agreement. If the clinical trial is necessary, it is precisely because recovery is not, in general, susceptible of bringing to agreement, of becoming the guarantor for a therapeutic technique that would, finally, have asked "the right question."

Correlatively, the agreement in question can be called into question, whether it is because it is a matter of broadening the convention to new protagonists who weren't associated with it,[26] or because, one of these days, the other protagonists will conclude that the pharmaceutical industry really does profit a bit too much from its role as judge and participant.

"Speaking well" of the process of pharmaceutical innovation is thus positively to exclude what I call "scientific practices" without, for all that, passing judgment. The sciences will not bring a response "finally free from any cultural belief" to healing, a question that interests everyone. Rather, it is the amalgam called "science" that allows doctors to present the cures that they obtain as "science-based" and to dismiss everything else as "placebo effects." The lie begins when the "placebo effect," a statistical category corresponding to "improvements not associated with the taking of the molecule that is being tested," is transformed into an operator for judgment against the so-called untested therapeutic practices. It then becomes a weapon of mass destruction, as blind as the statistical method from which it derives.

It must be emphasized here that it really is a question of a destructive weapon, because not only does the statistical method have strictly nothing to say with regard to therapeutic practices; it also implies a form of "antipractice," in the sense that it defines the question of the therapeutic relationship as an obstacle to get around. Of course, the patients in a clinical trial know that they are dealing with a doctor and behind him, all the promises of "progress," but this knowledge is "something that cannot be avoided," and everything that is not the neutral gesture "swallow this" is defined as producing "bias." In other words, the clinical trial is anything but "neutral": *it is in the service of therapeutic means that can claim to disregard the ensemble of practical knowledges cultivated by therapists.*

Can one imagine what a science that sought to address itself to the process of healing as such would be like? The first condition would certainly be that, instead of holding forth on the placebo effect, doctors transform the unexplained recoveries that they have all encountered—but which they often prohibit themselves from talking about, except in discrete and private confabs—into public knowledge. Another would perhaps be the refusal of any generality—such as the invocation of the sociopsychosomatic dimensions of illness and recovery—if this invocation doesn't communicate with any practice. The triple distinction invoked by the term "socio-psycho-somatic" effectively ratifies disciplinary categories—notably, it supposes that it really is "the body" that so-called somatic medicine takes charge of—and the associating of what has first been separated is mute with regard to the learning process that might start up.

As to this process, only one thing seems to me to be certain: the amalgam formed between the clinical trial and "science" is the worst advisor on the subject, because it blocks the question precisely where it should be asked. This amalgam effectively excludes the artisans of "untested" cures and supposes that "science" should be able to understand what they do better than they do themselves, and without them. The question of healing then remains defined by the convention that keeps them at arm's length.

CHAPTER FOUR

THE SCIENCES IN
THEIR MILIEUS

The Fable of Origins

Let's come back one last time to the particular group of theoretico-experimental scientific practices, because they are at the same time both what is at stake in the amalgam that we tried to undo in the previous chapter and its birthplace. At stake because "Science" is nothing other than the claim to make all the fields that present themselves as scientific benefit from the strength of these very particular practices. Which also means that the differences between these fields will be described in terms of distances from the ideal model that these practices constitute. But the birthplace, too, because this birth is marked by an event that everyone remembers and that many identify as a veritable advent: the "Galileo Affair." Despite his condemnation, Galileo Galilei's triumph over Rome supposedly constitutes an event that makes the difference between the past and the future of humanity in its entirety. And this novelty enables the amalgam because it is defined in the first place by what the "new science" had to confront: the authority that traditional knowledges (theology and philosophy) assumed for themselves with regard to the order of the world. After Galileo, the greatness and destiny of every science would be to confront a tradition, to deny what this tradition secures adherence to.

In order to pose the question of this event, which is supposed to mark the history of humanity, I will linger for a moment on the Galilean experimental event in the "abstract," stripped of what is supposed to have made it an event for everyone. My objective is not to deconstruct it but to underline, at the same time, both its effective novelty and its restricted scope, because the event in question does not bear on the astronomic, even cosmological, question of the relationship between the Earth and the Sun. Even if Galileo has left a particularly spectacular polemical mark on the history of heliocentrism, the thread of that history passes, in the first place, via his contemporary, Johannes Kepler.[1] Contrary to Galileo, Kepler dared to "break the circle" by attributing to the planets a movement that for him was an "unremarkable," or a "matter-of-fact," one, simply fitting with the observed facts: the elliptical movement that Newton would subsequently associate with a "central force." It cannot be emphasized enough that it was instead in terrestrial physics that Galileo made a difference: in a mode that was irreducible to any polemical effect but to which he owes his presence in physics textbooks. And more precisely, he made this difference in a physics that was "scarcely terrestrial," because the Galilean definition of the manner in which bodies fall is indissociable from the first "experimental apparatus," producing facts that were able to get their interpreters to agree, and because this apparatus implied the maximal exclusion of the most general form of "terrestrial solidarity": friction.

Also, it cannot be emphasized enough that this apparatus stages something that philosophers had, since Aristotle, agreed to judge as perfectly intelligible, without mystery: the movement of falling, the spontaneity of which provides so-called heavy bodies with their definition. It is this movement utterly lacking in mystery that Galileo interrogated in a new way in 1608: an inclined plane is set up on a table; a ball rolls down the plane onto the table and then falls to the ground. That, at least, is what can be deduced from a scribbled page found among Galileo's papers (folio 116f).[2] Aside from a drawing, this page features some columns of numbers indicating that for each starting height on the inclined plane, Galileo must have measured the distance between the edge of the table and point of impact of the ball on the ground. Now, from the point of view of what will eventually be Galilean physics, the relationship between these measurements has nothing to do with a merely factual relationship between two indifferent lengths. It is a bringing into relationship that verifies a "reason," a well-defined articulation between two aspects of a situation. The distance of the point of impact "measures" a velocity that would have been gained along the descent and then maintained in a uniform manner during the trajectory across the horizontal table, before continuing, still in a uniform manner and in the same direction, while the ball fell to the floor. It is because the apparatus is capable of verifying the fact that the velocity is effectively "the same" in its three roles (acquired at a given instant, when the ball passes from the inclined plane to the table, uniform during the horizontal movement that continues for as long as the ball is still on the table, entering into composition during the free fall movement toward the ground) that it constitutes the prototype for successful experimentation such as I have already characterized it.

Galileo's experimental apparatus verifies a "reason" in the etymological sense of *logos*, or *ratio*, but it doesn't give the reason for this relationship. This is the novelty that makes for the particularity of the event. Galileo can become the "spokesperson" for the manner in which bodies fall, but he would have nothing to discuss with philosophers, those who take an interest in reasons in the sense of a "why," because the apparatus is mute about the reason why bodies fall in this way and not otherwise. Or rather, if there is a reason, neither Galileo nor the philosophers of his epoch knew what it was. They must all, Galileo and philosophers alike, yield to the verdict of the experiment: agreement is imposed by the facts.

In *The Invention of Modern Science*, I characterized the event of this successful experiment as "the invention of the power to confer on things the power to confer on the experimenter the power to speak in their name."[3] Power thus intervenes three times and three times with a different signification.

"The invention of the power to confer . . ." refers to Galileo, although it is what the inclined plane is able to perform that he discovers: the transformation of the common evidence according to which (heavy) bodies fall into an "articulated fact," a fact defined in terms of variables, that is to say, magnitudes that Galileo could vary independently of one another, but which the fact articulated explicitly by proportional relations (and, after Galileo, by functions). In other words, the inclined plane creates the interreference between function and state of things that, for Gilles Deleuze and Félix Guattari in *What Is Philosophy?*, is the properly scientific mode of creation. If, in physics, this creation gives its singular power to mathematics, this power is not general. It requires the successful invention of an experimental apparatus allowing for the identification of and functional articulation between variables defining together the staged behavior.

". . . on things the power of conferring on the experimenter . . ." is the very definition of what is authorized by successful experimentation. It is what one might, with Deleuze and Guattari, call a "marriage against nature," a knot between two lineages that have nothing in common with each other. Before the event of this knot, falling bodies had only indifferent relations with the knowledge that we can produce of things. The fact that bodies fall had no power to oblige thinking, only that of illustrating the logical articulation between the general and the particular: all heavy bodies fall; this apple is a heavy body; therefore (if I let go of it), this apple will fall. After Galileo, the physicist would no longer deal with apples falling from a branch, with the crashing of avalanches down mountainsides, or human bodies falling out of windows, but with something new: "Galilean bodies," the privileged locus of which is no longer the center of the Earth, as it was for Aristotle, but rather the laboratory (and the sky), since their movement must maximally approximate the ideal of a frictionless body. A relationship was created that allowed Galilean bodies to testify, to play a determining role, in a human argument. Correlatively, the discursive line of argumentation will be linked to that of the successive roles that such bodies would play. Enrolled as Galilean bodies, the moon, the planets, and the comets will most notably have the power to authorize Newtonian astronomy to overturn the order of the knowable by introducing the incomprehensible "force of attraction at a distance."

". . . the power to speak in their name." Galileo represents falling bodies; he is their spokesperson. He no longer needs to argue and can turn his back on his human brothers, break the intersubjective debate in the search for common reasons. The experimental apparatus thus gave to Galilean bodies the power to allow Galileo to remain silent, to restrict himself to "showing the facts." But questions are then posed: "Whom will he address? Whom will he show them to?" Who will celebrate as an event the fact that Galileo now knows how to describe the movement of a falling body in a frictionless environment? This is where the question of what one can call a "social construction" is posed, because what now matters is the signification acquired by successful experimentation. A question that is all the more delicate, given that success here concerns the most common, the most predictable phenomenon one can imagine.

When physicists are really angry, they often propose that adversaries who accuse them of relativism throw themselves out of the window in order to see "if the law of falling bodies" is merely a social construction. This angry stupidity marks the final triumph of the "social construction" that Galileo, the first in a long series, was the initiator of: to succeed in acting as if physics demonstrates that heavy bodies, whether or not they think, and whatever they think, fall. A very particular success—we know how bodies fall, at least if, as a first approximation, friction can be excluded—is thus transformed into an event that should interest everyone and, first of all, those who forget, with ingratitude, that it is to physics that they owe their capacity to differentiate between doors and windows.

Galileo, who discovered the possibility of transforming rolling balls into "reliable witnesses" of the manner in which their movement must be interpreted, was also the inventor of what could be called the "territory of science." This invention is characterized by a rather extraordinary double movement. In the first place, it's a matter of chasing out of this territory all of those—philosophers and theologians—reputed to be intruders, in order that the new class of legitimate inhabitants may be defined by means of a contrast with them. The latter will be those brought together by successful experimentation as such: those I have called "practitioners," experimenters as much as theorists, who are obligated by the fact that such success is possible and who produce the innovative verifications that will also extend its scope (but then . . . ; and if . . .). But to be able to be "chased out," philosophers and theologians have to feel a concern. It is not appropriate that they shrug their shoulders, learning with polite indifference that it is now possible to determine how bodies fall, only to consider that, all in all, this doesn't matter very much for their own questions, so they may go on busying about. It is necessary for them to *feel that they have been chased out,* which is also to say that they remain fixed at the frontiers, fascinated by this new manner in which "their" questions are answered without them. And that they thus represent all other humans, whose questions have now been answered. Galileo is the inventor of an "epic" genre, in which the scientific hero, bearer of the future, expels the imposters who turn their backs on this future.

It is important to emphasize that this epic genre intervenes in the history of the sciences but is not constitutive of some continuous identity. One could say instead that as a genre it has constituted a resource for incessant repetition and notably for the resounding foundation of every "new science." The driver for this repetition is always the same: on the one hand, it's a matter of transforming the question that has been rendered decidable into a question that, properly formulated at last, interests "everyone"; and on the other, of identifying the "residue"—what the new science doesn't give an answer to—with what has until now been an object for irrational fascination or mere opinion. At one and the same time, this places the answer that has at last been obtained under the sign of a heroic combat against these monsters and permits the demand that everyone kiss good-bye to this residue: to be interested in it is to be suspected of irrationality.

The "Galileo Affair" is the prototype of this epic genre. Galileo could not "prove" that the Earth turns: the manner in which bodies fall only allowed him to suppress a crucial argument in favor of its immobility. But presenting himself as the spokesperson of an Earth that "really does" orbit the Sun, and not as the partisan of a "simple" hypothesis in astronomy, was well and truly to capture a question that interested everyone (the literate, in fact) to the benefit of his new science and to construct around this question the territory in which "the facts" are in command and from which the representatives of faith and discursive reason must be excluded. It is this first operation that is fully deployed with the publication in 1632 of the famous *Dialogue Concerning the Two Chief World Systems*, which would earn Galileo the anger of Rome. In effect, however hard Salviati—representing Galileo—tries to restrict himself (with fairly transparent hypocrisy) to the domain of hypotheses, he cannot prevent Sagredo—representing the cultivated public to whom Galileo was addressing himself—from noisily concluding that Galileo has proved that the Earth "really does" orbit the Sun.

But what is perhaps Galileo's greatest stroke of genius will come later. In the *Dialogue* the manner in which bodies fall was presented on the basis of arguments resting on unrealizable thought experiments. As Paul Feyerabend has quite rightly remarked in *Against Method*, it's a matter of persuading the public, of modifying its thinking habits, of ceasing to trust in the perceptual evidence, "I am immobile on an immobile ground!" By contrast, *The Discourses and Mathematical Demonstrations Relating to Two New Sciences*, put together after Galileo's conviction, constitutes a veritable testament, addressed to competent readers. The "third day," devoted to naturally accelerating movement, is presented here as a succession of demonstrations around the inclined plane: all the consequences of the success of 1608 are deployed there in an ordered, implacable manner. Galileo is addressing himself to his colleagues and is passing his knowledge on to them. But he also passes on to them a manner of making something, which would otherwise interest only them, "look good," get a general, crucial value.

Who hasn't sat behind a desk at school and heard that the sciences answer the question "how?" and do not enter into the "why" of things?[4] As if, for most of the questions that matter, the why and the how could be disjoined! A disjunction of this sort is not given; it is nothing other than the very signature of successful experimentation. Its transformation into an epistemological generality is thus the negation of the event that this success constitutes. But in the latter's place comes an event that for its part interests humanity as a whole: an undertaking that is "scientific at last," which has been able modestly to restrict itself to the "how" and which has reaped the benefits of its modesty. This is the style of presentation that Galileo initiated at the moment he announced that he could define the properties of the accelerating movement of falling bodies without having to inquire into the cause of this acceleration. Because he doesn't comment on the novelty of the experimental success as such, he acts as if the disjunction between the how and the why preexisted, and he dismisses the question of "why bodies fall in this way" as a sterile debate among philosophers, "imaginations" that there would be little benefit in examining.

In other words, Galileo cannot respond to the kind of question that, until then, had been what concerned thinkers, but he defines it as that which can only interest lovers of interminable discussion (or "conversation," as Richard Rorty might put it). What can be decided, on the other hand, is the cause, in the sense that this cause is caught up in an articulation between cause and effect, in which the one measures the other and vice versa.[5] Those who accept the division between why and how as an honorable distinction and claim responsibility for the riches and unfathomable uncertainties of the why fall into a trap that is therefore already set. Because from the point of view of the scientific "how," the "why" is only the (as yet) undecided residue of what it has been possible to decide, a residue that has no other consistency than the "bad" questions that it has provoked. Actually, it is indeed a residue without any consistency, as it will be possible to redefine it incessantly as a function of the "progress of scientific knowledge." From the point of view initiated by Galileo, there is no symmetry: the adepts of the "why" will have to be ready to give up the terrain that is conquered, even if only rhetorically, by the "how." And correlatively, once a proposition is able to profit from taking up the disjunction so as to install itself on the separation that has at last been accepted between the "facts" that permit a "how?" to be answered and the "why" subject to conflicts of value, it will benefit from the evidence associated with the progress of rationality.

But the most formidable of Galileo's inventions is the explicit mobilization of the thesis according to which every definition is the arbitrary work of the human, elaborated in the abstract. His stroke of genius was to have presented this thesis as an objection, an objection that doesn't come from Simplicio—assigned the task of representing all Galileo's adversaries and of suffering all the slights—but from Sagredo—representing the enlightened public, always convinced by Salviati's arguments. Sagredo remarks, in a very respectful manner, that it is difficult for him to accept that the movement for which Salviati announces the definition is the "natural" movement of heavy bodies. It is effectively a matter of a definition that is elaborated and accepted in the abstract, and such definitions, however they are authored, are arbitrary. Now, Salviati accepts the general legitimacy of this objection, but, thanks to the inclined plane, he shows that the definition of uniformly accelerating movement *is an exception.*

Galileo thus created a solidarity between the definition that is "scientific, at last" and a radical skepticism inherited from medieval nominalism. It is in effect in relation to this skepticism, *the legitimacy of which he recognizes,* that his "new science" will be an exception, an exception that will concern everyone who wishes to escape from arbitrary interpretations. This signifies that the new, scientific, territory found the means to define its environment in a single stroke, not by means of contrasts to be negotiated but through a binary judgment: "us" versus all those others who claim a knowledge of things. These others may well be different from one another, but from "our" point of view, they must all be lumped together in the same bag, that of a fiction that entertains only arbitrary relations with the facts. As for us, we know that reason is reason only by accepting that it has no choice but to follow facts.[6]

Evidently this is not history in the strict sense, in the sense wherein I would respect the rich complexity of Galileo's position, like that of the political, religious, and cultural crisis that marked his epoch and shook the legitimacy of all authority. It's a matter here of emphasizing Galileo's stroke of genius: his argument implicates this crisis in a mode that doesn't just place his science "outside" the crisis but turns the crisis into a figure for the new authority. Successful experimentation is transformed into the ahistorical foundation of a theory of knowledge that skeptically dismisses everything that cannot be decided scientifically. At the start of the third day of the *The Discourses and Mathematical Demonstrations Relating to Two New Sciences,* there thus appears—in its native state, one might say—a mode of presentation that, far from having aged, has today gained the status of common self-evidence.

Moreover, this common self-evidence also defines constructivist sociologists, who testify twice over to the efficacy of Galileo's staging: by wanting to show that nothing can form an exception to the skepticism concerning every theory bearing on things and by refusing to enter into scientists'"reasons," into their "why," so as to limit themselves to a (sociological) "how" that they would like to be "purely factual." What had scandalized those in Galileo's epoch for whom reason referred to the search for reasons has thus become a habit of thought. Galileo has "won"; critical sociologists are his inheritors too.

I will thus propose that what has been called "Science"—that which restricts itself ascetically to the "how" of things, which wins a knowledge that is finally valid where only arbitrary opinions previously existed, where conflicting chatterboxes predominated—is nothing other than an "interface product." "Science" cannot define itself independently of its foil, independently of a milieu that accepts as its self-image the ensemble of judgments that are organized around the qualifier "nonscientific." This image draws its power from the manner in which it captures and reassembles the questions and fears that belong to this milieu, for its own benefit, transforming them into a stable definition of the milieu of every science. Neither Galileo nor Robert Boyle, who would later codify the ethics of gentlemen experimenters, invented skepticism, the question of the authority able to transcend conflict, or the appeal to facts in the face of human imagination. But "Science," born in an epoch when these questions had acquired a critical importance, "created" an event by transforming them into operators in the definition of its own identity.

The Construction of a Territory

The epoch that resonated with the novelty of the public announcement that, thanks to "a fact," humans had found the means to overthrow a knowledge supported by a thousand rhetoricians and a thousand and one reasons is long distant. And without a doubt what created the most powerful temporal distance is the transformation, in the nineteenth century, that simultaneously saw the appearance of the modern state, the institutionalization of scientific research practices in the university, and the spectacular, spectacularizing, link between this research, techno-industrial progress, and human progress in general. What I would like to comment on here is not the success of the institutionalized scientific undertaking but a new problem, linked to this change of milieu. Scientific territories would henceforth demand not only that the representatives of the "old questions" serving as their "foils" both be chased out and become fascinated but also that new protagonists—who are not fascinated but perhaps a bit too interested—be kept at a distance. Scientists are on the defensive, but they cannot "chase" these new protagonists away, because the financing of their research, like the "worldly" extension that permits scientific progress to be identified with human progress, depends on them.

It's not a matter here of doing the history of this transformation, that is to say, of establishing what Dominique Pestre calls a new "regime of knowledge." It is a matter of giving what I am calling "requirements"— what scientific practitioners require of whomever or whatever they are dealing with—all their importance, and of distinguishing "requirements" from what I have called "obligations." As it happens, such a distinction has to be *made*, not recognized as if it were inscribed in states of affairs themselves. Making this distinction means protecting it from the idea that it is a matter of two sides of the same coin, something that would amount to turning obligations into an identity requiring recognition: we are obligated as scientists and by virtue of our "scientific mind," and hence we require.

So it's the requirement of "autonomy" that is going to pose a problem here. This doesn't signify that Galileo or Robert Boyle were "naturally autonomous." Certainly Boyle was a rich nobleman, and thus independent, but contemporary historians have shown us Galileo as a "courtier," constructing a network of support and protection in the courts of the Medicis and the pope. As for the academies of the eighteenth century, they depended on the goodwill of their patron, the enlightened sovereign. What is new is the disquiet, regarding state and industry imposing their priorities, for which the German chemist Justus von Liebig becomes the spokesperson in 1863, with his *Lord Bacon*. Liebig knew much better than Galileo just how many apparently "good" questions, apparently well-established facts, his science had to disregard. However, it wasn't possible to treat them as a residue, to disqualify the interest that they stimulated by characterizing it as "irrational," because this interest came from protagonists who would not allow themselves to be impressed. The experimenter now had to defend the relative rarity of "experimental facts" in the face of the rising tide of interesting, reproducible, facts.

"We require" poses problems that are very different from the question "to what does this situation oblige us?" The first difference results from the contrast between the generality of the requirement and the hesitant, always circumstantial and imaginative character of the question that unites practitioners. Even the requirements that condition successful experimentation are of the order of a generality. That it must be possible actively to stage what must be interrogated without it being reduced to an "artifact," that is to say, without it being possible to attribute responsibility for the response to the staging, is a general imperative that is mute with regard to "how" it will be satisfied. The requirement of having to avoid artifacts communicates with the sounding of an alarm that every experimenter understands the meaning of. But it pertains to every field, in every epoch, in every new situation, to produce the means for claiming to have satisfied it. But the generality of requirements finds its acme when it is a matter of those requirements that the "social" environment of a science must satisfy. It can then very easily turn into an "order-word" and mobilize a varied ensemble of innovative or hackneyed slogans to do so. Thus taking up the theme of faith for themselves, in order to require that this faith not be shaken by the demand for accounts involving a lucidity that would kill it off, was a bold move on the part of certain physicists at the start of the twentieth century. But today, the faith that is indispensable for physicists has become a classic theme. To plead the case for the "large-scale equipment" needed by high-energy physics, some have even grafted onto this faith a reference to medieval cathedrals, as a manifestation of faith but also a center of economic life. The argument for "chance discovery" has also been much used, with examples at the ready, in order to demonstrate the harmful character of programmed research—although research programs in biotechnology have not, for their part, hesitated to announce dazzling progress in the domains of health, the protection of the environment, and the fight against hunger, as if considering it done. And again and again the image of a science demanding protection from opinion, imprisoned in its fictions and irrational fears, is mobilized—although now it is rather a matter of defending oneself against the interests of protagonists whose rationality can no longer be called into question, because they provide for research. Grappling with those whose interest in them is threatening, scientists most often continue to flatter them, the way one flatters one's patrons, and they ask them not to listen to opinion, which, freed from science, would kill the goose that lays the golden eggs.

Whereas obligations make practitioners hesitate, the arguments that justify their requirements are marked by a kind of conformist indifference, as if everyone could draw from a box of "ready-made" platitudes without the slightest concern for coherence. Thus the majority of scientists will not hesitate to chalk up to the progress of science whatever seems to move in the direction of consensual progress, but whenever the "application" is contestable, they will use the image of the axe, which is not responsible for the murders it is the instrument of.

It will have been understood here that the respect that I am pleading for with regard to the obligations of practitioners does not correspond to the necessity of any kind of respect at all for the manner in which scientific communities demand respect, in the sense this time of the "respectable distance" at which they hold those whom they nonetheless need. But this doesn't, for all that, mean that obligation and requirement can be opposed as one customarily opposes scientific validity and ideological claim. Every practice needs an environment that agrees to provide for it without enslaving it. The manner in which a practice translates this need into a requirement and constructs its own identity on terms that must incite its environment to satisfy these requirements implies the question of this environment itself, that is to say, of what it is ready to accept. This doesn't mean that scientists could, from one day to the next, produce other, more interesting, more lucid arguments, if the milieu was sensitive to them. In the case of GMOs, when the usual arguments were suddenly struck with impotence (the more they were used, the less popular they became), there was profound disarray. It was a bit as if a shepherd, whose skills match with "what counts" for sheep, was suddenly dealing with a herd of pigs, for which those skills no longer worked. But the shepherd can learn, and so can scientists.

Rather than talking of ideology, one might say that the requirements of scientists refer to the epoch of which they are an integral part, in the same way as those to whom they are addressed. This was already the case with the "fable of origins," that is to say, with the manner in which Galileo captured and benefited from the crisis of his epoch with regard to the legitimacy of authority. In any case, he himself was doubtless the first to vibrate at what he had just discovered: "There are facts that are 'authoritative'!" It was also the case when scientists working in the academic milieu accepted what has now become a habit but which might not have gone without saying: that scientists working in the private sector (or for the military) are defined by the interests that they serve. That is to say, that they are released from any other obligation and assigned the task of transforming into the golden eggs of innovation, that which the goose of "fundamental" research continues to produce in all innocence. A division of labor of this kind is effectively limited to a rather servile translation of the assumed hierarchy of wage earners and liberal professions. This was the case once again when, after the Second World War, American academic research experienced its Golden Age with the taking up again by the state of the grand theme of fundamental research updated for the Cold War and anti-Communism. The scientists who would take up the arguments that present "Science" as transcending political conflict belonged to an epoch in which Nazi fanaticism and Stalinist terror were associated with a triumph of ideology, against which Science alone was the bulwark. Fundamentals would imply, primarily, being a stranger to political questions, but also being the source of an industrial progress destined to short-circuit political questions. Who is interested in how the pie is divided up if it never stops getting bigger?

The question of the requirements of scientists must be distinguished from that of their obligations because what a science requires belongs to an epoch and is therefore the locus of an unknown that designates the future: What new formulations might scientists be able to give to their requirements? However, this future, our future, might equally see the destruction of scientific obligations. The position taken by Dominique Pestre in *Science, argent et politique: Un essai d'interprétation* is particularly interesting from this point of view. As we have seen, as a historian he describes a succession of regimes of knowledge and underlines just how mendaciously the regime that is in the process of being destroyed presented itself. As the so-called goose that lays the golden eggs, scientists ostensibly preoccupied solely with the advance of their science—requiring that they be left in peace, with the faith that would allow them to move the mountains that were commonly held to be immovable—were, in fact, in the process of cementing some very tight bonds with industry. The goose masked a strategist. However, Pestre does not conclude from this lie that the new regime of knowledge being put in place will simply be the one that "follows." He is worried, even if these worries are difficult to articulate with the prudence a historian needs, given the apocalyptic visions that every epoch in crisis provokes. And it is the distinction between requirement and obligation that allows me to relay these worries.

To characterize the regime of knowledge that is in the process of being dismantled, I will refer to the model presented by Bruno Latour, originally in the form of a fictional diary describing the week of a research laboratory "boss,"[7] and then in the more systematic form of a description of the "vascularization of facts," that is, of the circulatory apparatus that gives life and value to facts, conferring on the new thing that a science is in the process of learning its "innovative" character, a vector for socio-industrial innovation and for the reorganization of scientific disciplines.[8] In other words, this model allows the posing of the question of knowing how the eggs that the goose lays become gold: not because they are harvested by interested protagonists who would come and collect them from under an indifferent goose, but because these protagonists have been actively interested by what interests the scientist. Here, then, we are leaving the question of the "lie" constituted by the presentation of the goose that lays the golden eggs, that of a "disinterested" science whose "benefits" are then celebrated, in order to characterize the feat that is masked by this lie.

The innovative scientist—Pierre Joliot-Curie, in the case studied by Latour—is in no way a sleepwalking scientist, asking only to be left to work in peace about him. He circulates incessantly, entering relations with different interlocutors, with divergent interests. He has to be "in action" in multiple, distinct ways if his eggs are to have a chance of becoming golden.

According to the specifications constructed by Latour, the innovative scientist needs "alliances." He needs nonscientific protagonists who count, who have the means, to agree that what is promised could be worth their support. If the scientist doesn't manage to recruit such allies, he will have to make do with the inadequate funding of "purely academic" research. But he must also "mobilize the world," that is to say, equip himself with an ensemble of resources, instruments, equipment, without which his "ideas" will remain ideas, with no hold on what they target, and his allies will betray him. He must also ensure the "autonomy" of his undertaking within the scientific community and have its credibility and the specific training corresponding to it, its own criteria for relevance and validity acknowledged. Without this, his work would not belong to the grand history of scientific breakthroughs; he would not be able to attract brilliant and ambitious students, nor propose "facts" that his colleagues will have to take into account. And finally he must construct the "public representation" of his research and what it aims to do: he must present this research in such a way that the public recognizes its legitimacy, in the service of progress and of disinterested knowledge.

What's at issue here are so many vital requirements and the fact that just one of them might not be satisfied risks jeopardizing the undertaking as a whole. Thus American (very) high-energy physicists have never been able to recruit strong economic allies and were dropped by their political allies when, in 1993, the US Senate refused to vote in favor of the funding necessary for the extremely costly equipment mobilization that they judged to be necessary. Thus also, in spite of the weight of their allies and the professional recognition of their specialists, GMO crops collided with a public representation that became more and more disastrous, as the allies with whom and the interests around which they came into coalition were identified. As for Jacques Benveniste and his work on water memory, it was the question of autonomy, that is, of the recognition of the legitimacy of his undertaking by scientists belonging to other fields, on which his defeat turned, allowing his work to be dismissed through reference to the interests of his ally, the homeopathic medicines industry.

The distinction between the four types of protagonists that an innovative science has to interest ensures a form of autonomy for this science, one that scientists today think was a matter of a "normal" situation (in the process of being destroyed). In effect, none of the "external" protagonists alone suffices, that is, none can play the role assigned to another, and that is what keeps each of them at a safe distance. Most notably an ally's interest does not allow the necessary recognition by the community to be short-circuited, except in unanimously denounced pathological cases (Lysenko!)

However, nothing seems opposed to the conclusion that this montage is anything other than a "social construction," the complexity of which doesn't prevent it being reduced to a set of negotiations and conventions between humans. But in reducing it like this, the meaning of the term *construction* changes. Construction here doesn't follow the contours of a preexisting distribution of interests; it is irreducible to a question of "influence," in the sense of being subject to influence. The scientist described by Latour has not been influenced: he is active; he creates "with" those who, for their own reasons, can be led to give a new scope, a new importance, to what he does. He is, like the Galileo of the *Discourse*, a speculative entrepreneur: it is the scientist who endeavors to influence, to confer on his proposition the significations that are likely to stimulate and recruit the interests that he needs. In other words, he is working at the interface between "his" territory and an environment that is, a priori, indifferent, and which it is a matter of transforming into "his milieu." Or, to be more precise, into his milieus, because the mode of presentation-proposition of the scientist can differ, depending on the interlocutor, capturing and translating to what should interest the type of interlocutor whom he is addressing.

Latour has never stopped emphasizing that the scientist characterized in this way is an entrepreneur who cannot be "judged" in terms of social relations that would explain his success, because he is himself the active, enterprising cocreator of the "social" on which he depends. Latour thus asks that his sociologist colleagues learn to follow scientists, to describe how a scientist makes an ensemble of heterogeneous interests "hold together," how she unites them around "this science that is important to us all," without their being united by any common definition of this importance. What is more, sociologists will have to follow a very particular kind of entrepreneur, because there is a fifth regime of activity, on which the other four depend, which it nourishes and is nourished by them: what Latour calls the "bond" or the "knot."

Allies, resources, professional recognition, and the respect of the public are not enough. All of Joliot's work of connection would collapse if the neutrons did not respond to what he expects of them. The neutrons' response certainly owes its importance to all his activity of "social construction," but this construction as a whole only makes the role that he does or does not succeed in getting the neutrons to endorse more important. The scientist who dances in his laboratory is not seized by the pure, disinterested joy of knowledge; every component of his success dances with him. But what makes him dance is not the article that appears in a prestigious journal, nor is it the research grant; it is the event of a success that needs everything else but cannot be determined by it. This success, which ties Joliot's innovation to the neutrons, also links him to his "competent colleagues," to practitioners who, like him, are obliged by the difference between success and failure that puts them into suspense.

Not to Stop at the Lie

The fable about the goose that lays the golden eggs is a lie. One can, in all tranquility, affirm that a scientific territory—what unites practitioners— is not "autonomous" in the sense that it would find both the reason for and the law of its development in itself. As for what is called the internal history of a discipline, narrating how it is that questions, experimental discoveries, hypotheses, and verifications alone suffice to account for the progress of the "content" of a discipline, it is an a posteriori construction that is, moreover, involved in the "public representation" that a science proposes of itself. But can one, for all that, say that this construct is itself also "mendacious," in the sense that it would be able to continue undisturbed even if the new regime of knowledge smashes the myth of the autonomy of scientific practices to pieces? In other words, is autonomy merely a myth meant to impose respect and submission? The notion of practice permits us to inhabit the slight difference between the lie that justifies autonomy and the importance of differentiating the scientific territory.

To envisage the situation from the point of view of practices is, first, to recognize that the differentiation between the inside and outside of this territory is permanently under construction by scientists. To recognize this, however, does not imply that this difference can, as result, be reduced to a simple illusion. On the contrary: one can say that the strength of this scientific construction is to make exist together, actively, strategically, and in a creative manner, a difference between what crucially matters for scientists, what obliges them to hesitate and might make them dance, and what belongs to shared history. It is this difference that the classic internalist history of the sciences will subsequently translate as "text" and "context" but that only ratifies what has already been prepared by the manner in which scientists produce and try to get recognition for the frontier between what really counts and what is a concern for a history marked by contingency.

A frontier like this this could certainly be assimilated rather easily to a convention, but this assimilation is precisely what the notion of practice proposes to resist. A convention is something that is agreed between protagonists who have an interest in it but could do without it. But it is vital for scientists that their allies do not interfere directly in the production of the links and knots that unite and obligate them, that these allies don't resolve their hesitation by imposing a criterion for success, even if they certainly have many means to help with the decision. Those who refer to these means so as to deride the claim of scientists according to which their allies don't weigh directly on the decision are acting as enemies. What the idea of a simple convention risks rendering "normal," indeed legitimizing,[9] is what scientists—and rightly so, from the point of view that I am defending—designate as pathological. That is to say, a history in which the alliance with an industry would, for example, make it possible to forego a controversy with colleagues, or in which a concern for public respectability would silence certain competent objections. A denial of the obligations that link scientists.[10]

Certainly the construction of the frontier that differentiates is far from being neutral: no construction is neutral. But rather than calling it mendacious or arbitrary (it's "nothing more than" a construction), it's worth distinguishing between the requirement to differentiate to which it responds and the manner in which this requirement is formulated and put to work. It's the manner in which the requirement is formulated and the means by which it is satisfied that can be assimilated to a convention, implicating protagonists with divergent interests. In our case, one singularity of the convention is that "nonscientific" protagonists agree to respect a mode of presentation that is silent about their own, active, role, their own interests, which makes it as if the "support" obtained by a scientist was a neutral condition, responding to the "quality" of her research, and as if the "effects" of this research came about only after its autonomous and disinterested verification.

It's not surprising that the scientific community in the broad sense, on which the recognition of the autonomy and credibility of an innovative field of research depends, respects this convention: reference to autonomous, disinterested development, which any constraint would destroy (the goose that lays the golden eggs) is what gathers the community together. This reference constitutes the kind of protection that some scientists have laid claim to for more than a century, and it has become today a veritable order-word.[11] But it's also the possibility of "mobilizing" the world that is at stake, because these possibilities vary and respond to different conditions, depending on whether research presents itself as responding to the imperatives of disinterested knowledge or as applied, in the service of "particular" interests. This is why maintaining a difference between so-called public research, financed by the collective, and private research, is also at the center of the preoccupations of the "allies." It is the stake of multiple assemblages all safeguarding the possibility of staging a science that is autonomous, in control of what can be represented publicly as "purely scientific progress," the source of a knowledge finally free of conflict. A public representation such as this matters for all protagonists equally because the "effects" of this science will also be able to claim to transcend conflicts: science as the brains of the progress of humanity.

We will call the ensemble of protagonists described by Latour who have an interest in respecting the convention, "active." For its part, the "public," which is the target of the "public representation" of a science, is supposed to adhere to what the convention stages. It must, at one and the same time, be situated by the convention and "outside" it—kept at a distance from the convention as such, not party to it but nevertheless supposed to "believe" what the convention claims. The possibility that the confidence of the public might be shaken corresponds to the scandal that brought about the "science wars." From the point of view of active protagonists, it doesn't matter very much whether those who are accused of having attacked the public's trust in "its" science defined the sciences as a "social construction like any other" or, like Bruno Latour, tried to follow its specificity and to characterize its strength. Either way, by trying to "make public" something that ought not to be, they have betrayed "Science."

There should be no need to underline the political character of this public representation, which continues to prolong the Galilean fable of "Science" as the brains, at last in action, of humanity. This fable now interests, first and foremost, the "allies" on which scientific communities depend: state and industry. Because to keep the public "at a safe distance" is also to keep that which is too important to be left to politics from being questioned.[12] And from this point of view it is definitely interesting that politicians, like scientists, believe themselves to be separated by the unbridgeable abyss between facts and values and that scientific experts are inhabited by the disjunction between the scientific, or rational, point of view and the subjective point of view, which isn't validated by what Americans call *sound science* (a science that is "tried and tested"). A decision based on tried and tested science has to be accepted as much by politicians as by the public, because politicians, just like the public, have been captured by the setup of scientific territory.

We can obviously say that this aspect of the setup—what it keeps at a distance—also serves the interest of scientists directly, as it allows them to prolong into the present the staging that opposes the scientific point of view and everything else, which is identified with human arbitrariness. But it's here, precisely, that the difference can be ventured between scientists as practitioners and scientists insofar as their identity, the manner in which they present themselves but also represent themselves "to themselves," is coproduced with their milieus. It must never be forgotten that, like everyone else, scientists were also young once and that, like everyone else, they learned in the classroom that Science is opposed to opinion, that objectivity imposed itself against the beliefs or irrationality that human subjectivity makes prevail.[13] Each of us, within our own specialties, has subsequently learned that things are a bit more complicated, and many might even, especially if they have had the chance to work in a "living" scientific field, regret not having been able to share their science "as it is actually done" with "the public." In other words, the scientist does not generally have an active interest in keeping the public at a distance. Contrary to other protagonists, she might not fear the public meddling with what concerns it by rights, but is rather convinced that, unfortunately, the public is incapable of meddling with science judiciously.

This difference, which foregrounds that scientists are also, in their own manner, "passive," adhering to the image of a public that must be enlightened, that is incapable of recognizing the difference between construction and fiction, is both a bet and a manner of addressing scientists without insulting them, at the same time. What is in question here is not their strength but their weakness, the fact that it is only on the ground, when they have become researchers, when they have themselves had access to science "as it is actually done," and this without cultivating the words to speak of it. In other words, the enlightening words that they have learned about science, at school and elsewhere, continue to inhabit them when they think about "science in practice," even if they know perfectly well that these words say nothing of what matters to them when they are working.

Ecological Crisis

The public representation of the sciences is not a simple ideology, a form of illusion that can be separated from everything else. It doesn't simply translate the manner in which the sciences are usually "presented" to those whose specific interests "don't count," that is to say, don't have to be recruited. It also concerns the manner in which scientists identify themselves, the manner in which they situate themselves and evaluate the relations that they are making incessantly with protagonists who have the power to confer on their propositions a signification "outside of science." The leitmotif is well known: science is "neutral"; it cannot be held responsible for those "choices" about its use that are judged negatively. It can sometimes happen that an individual scientist, like Einstein with the atom bomb, feels responsible, but that is usually in the mode of tragedy, that is to say, in the drama of science and the fate of humanity. On a day-to-day basis, scientists are not unaware that what they propose may entail "social risks," but they define these risks as the price of a progress that it is not for them to decide to deprive society of: they must limit themselves to hoping that society will be capable of managing innovation in an adequate and responsible manner. To do more would be to assume an undemocratic role in "society's choices."

What's at issue here is not in the slightest a naïve innocence proper to the scientific mind but something that the triple identity of Science—as motor of human progress, as direct and anonymous translation of a finally rational mode knowledge, and as producer of what cannot but be opposed to opinion, to inertia, to habits and traditional values—prepares and stabilizes. This identity might be associated with what Deleuze and Guattari describe when they show how a war machine can be "encasted":[14] the scientists of the Golden Age found themselves linked to and by the state and capitalism in a mode that nonetheless left them free to give themselves their own rules. They may have thought that they were benefiting from a "functional" or "logical" situation, that all were in their correct place, that the state and capitalism had understood the legitimacy of their requirements, and that the possible resistance of the "public" could only be explained by its ignorance and/or its irrationality.

But the weapons used by science criticism ratify this idea of science's function simply by interpreting it differently. The activity of scientists would, on this count, be a function of the interests and relations of force that prevail in this world, in relation to which scientists believe themselves to be autonomous. The struggle would therefore be against the privileges that a caste lays claim to and of making them accept their common lot. That these privileges are being called into question today, that the "allies" have apparently decided that there was no need to wait, to keep their distance, to abstain from interfering in the production of links and knots, is thus a nonevent: in any case, the distinction between internal and external histories, which scientists are so attached to, is a mendacious construction.

In order to stage the contemporary "betrayal" of scientists by their allies as an event, and not just as a "transformation" of a regime of knowledge, it is important not to assimilate such a regime to a "logical" or "functional" situation or to a "system" that would confer an identity on its terms and that will confer on them a new identity if that functioning is modified. I will associate the regime of knowledge that is in the process of being dismembered not with a "functional" system but rather with an *ecological assemblage* that associates heterogeneous protagonists pursuing divergent interests, united by relations that are not symmetrical, all protagonists making what unites them matter differently.

To speak of an ecology and not of a functional system in which every term plays a role that gives it its identity is to make matter the unknown of a future in which the urgencies of a "knowledge economy" would have swept away what unites scientists, what obliges them to think and to hesitate. A future in which the distinction between internal and external histories would become a merely mendacious construction. In fact this is already the case with the notion of the "gene," for example: the current history of the biology of the gene, which interrogates the stability of this notion without making the question of patents that need to link a gene to a function intervene, is simple propaganda.

To describe in terms of ecological associations the transition that we are witnessing—from an assemblage in which sciences are "encasted" (with all the privileges reserved for a caste) to an assemblage in which they are in the service of a "knowledge economy"—is to recall that such associations can mutate in a mode that is not a "response" to new, general conditions. An ecological mutation is not a manner of preserving "the system" through a modification of its functioning. It can also result in the destruction of some of its protagonists. In this case, to be sure, no one can yet foresee the repercussions of the mutation that is underway, but it could easily entail the generalization of the destiny previously only assigned to scientists working in the "private sector."[15]

The term *repercussion* is suitable for designating the possible consequences of this transformation. In a situation of an ecological type, it signals that no protagonist benefits from a relationship of mastery of the consequences, even if that protagonist is in a position to take the initiative. In this case, the contemporary statist-capitalist bet, according to which the "allies" of science can directly pilot the scientific territories judged to be innovative and can organize what they call a "knowledge economy," without for all that, "killing the goose that lays the golden eggs" is far from certain. Because it is not a convention that is betrayed here. On the contrary, the convention is maintained because research is still presented as being "in the hands of the scientists," scientists who would finally have agreed to leave the "ivory tower" and have discovered the benefits of an industrial partnership. What has been betrayed is the kind of keeping at a distance that successful science perhaps requires. The repercussions of this taking in hand, destroying what links scientists, the obligations that makes them practitioners, could easily destroy what makes them reliable partners for innovation. It may be that financialized capitalism has discovered that speculative promises can work as well as real eggs when it is a matter of their weight in gold.[16]

But there is another aspect to the contemporary "ecological mutation." The bet according to which the taking in hand of science by its "allies" can leave intact the "public representation" of science, which keeps the "public" at a respectful distance and keeps science "out of politics," is already proving risky. To be sure the "public," in the normal sense of the term, remains in its place, but such norms only mean something in a stable situation, whereas we are in a turbulent period, marked by the entrance on stage of groups who are learning to extricate themselves from the role of grateful beneficiary assigned to them, and they are doing so in a rather novel, ecologically significant, mode.[17]

That a part of the public "resists" is not, in and of itself, new. It happened in relation to nuclear energy and weapons and to the "military-industrial complex," for example. But what is unprecedented in the dynamic of the contemporary questioning of science is that it does not remain localized around a "cause," which it feeds by learning reasons, old and new, for interfering with what was supposed not to concern the incompetent. Correlatively, mediators usually tasked with "reassuring the public" have themselves started to stammer, as they cannot ignore the fact that the grand discourses on Science as the motor of progress that perpetuate themselves "at the highest level" might no longer "take anyone in."

It is therefore in relation to a situation of ecological "crisis," a destabilization with repercussions everywhere, even catching those who triggered it unawares, that the notion of "practice" may be relevant.

Until now scientists have been well served by the public representation of the sciences. But the notion of "practice" might make them think, because the representation of the sciences is now being turned back against them. They cannot make what they are undergoing public, because they cannot appeal to those who have been restricted to the double position of being interested spectators, who learn the right answers to their questions from the sciences, on the one hand, and beneficiaries satisfied with the by-products of scientific progress, on the other. Those who, like me, have wanted to offer words to scientists that are different from all-terrain words like *rationality, objectivity, scientific mind,* so as to be able to share the fact that the reliability of scientific statements can be destroyed if economic imperatives prevail over the obligations that link scientists have collided with this fear: if we tell "them" about science "as it is made," they will lose trust, and if they lose their trust in us, nothing will protect them from irrationality. When talking to children, one must know how to lie or to embellish, and adults must never, ever, argue in front of them!

However, a characterization of scientists as "practitioners" isn't primarily a "better representation," one that is more faithful and susceptible to being shared. It also corresponds to the possibility of a different ecology, because it no longer puts the emphasis on a knowledge that is intrinsically objective and reliable, that is "all-terrain," but on a success the reliability of which refers to the very particular milieu in which it has been produced. This signifies that reliability is not a stable attribute and is not in the least bit guaranteed when there is a change of milieu. It is thus a matter not only of proposing to scientists that they say farewell to the Galilean fable—which gave the possibility of defining how heavy bodies fall in a frictionless environment all the allure of an epic—but above all of inciting them to envisage different relations with their milieus. To envisage that they might be "obligated" to do this by the question of "how what is produced in their laboratories leaves their laboratories."

THE SCIENCES IN THEIR MILIEUS

It really is a matter of an obligation here, and no longer one of requirements. Because if the obligation is what makes for hesitation, because it implies that something that obeys no requirement "agrees" to the staged rendezvous, then maintaining the reliability of what leaves its milieu of origin does indeed depend on new types of rendezvous, organized around other kinds of questions. Maintaining reliability effectively implies the activation of questions and objections inhabiting the milieu in which the novelty coming from elsewhere is to be implanted. If the reliability of what is to be implanted is to be maintained, that is to say, re-created, then it is these questions and objections that have to be taken into account with the same seriousness, and probably more risks.

The conventional arrangement that is in the process of being dismantled implied the systematic rarefaction of what was entitled to serve as an objection, and, in this sense, it constituted a recipe guaranteeing a loss of reliability of the "worldly successes" of the sciences, a *not* treating the consequences that these so-called successes would entail as a demanding matter of concern. That the theme of "sustainable development" can appear as a new idea clearly shows that "sustainability" was the least of the worries of those who profited from this success. But for scientists, it is not enough to acknowledge that they played an active role in this recipe by ratifying in their own way the brutal differentiation between those who have the power to object and those reputed to be irrational. They must learn to make an active difference, a *matter of concern*, between "impact"—to have been able to interest protagonists who are able to make what they have produced leave the laboratory—and "success"—to have participated in a genuine relay, reinventing the signification of what they have produced in relation with what is not just "outside the laboratory" but also a new milieu posing different questions.

The ecology that the notion of practice appeals to is not simply "democratic," in the sense that this term supposes the general notion of the "public," citizens who are called on as such so as to have their own say. That "citizen juries" or "citizen assemblies" might take on a real importance would be a political event, the repercussions of which would certainly be profound and multiple for scientists and their allies. But what matters from the point of view of the ecology of practices is the manner in which problems might be deployed. For scientists to become capable of actively participating in this deployment and not just putting up with it, they must become capable of different relations with their practice. In effect, it cannot be a matter of retaining and trying to reconcile two terms, which only have a precise meaning within the network of oppositions constructed by scientists themselves: the facts on one side, the values with regard to which it pertains to the "public" to make up its mind on the other. It's a matter of abandoning the Galilean fable that takes the place of "political conscience" for too many scientists, the appeal to "real democratic debate" to decide what concerns "everything else" once the "facts" have been established— what today is often called "ethics."

The ecological approach thus imposes its own requirements here; it asks that the ensemble of protagonists interested in or threatened by an innovation with a scientific origin be put on the same plane; it asks that they be recognized as heterogeneous and not situated in a way that conserves the fable of origins, with everything that is not "scientific" being put in the same (ethical) sack. Correlatively it is not certain that the central character of the categories of convention and objection can be retained, supposing the gathering around a common question in the strong sense—that of an interest that is shared, albeit in different ways, by all those gathered together. The question of what protagonists make matter, what makes them hesitate, what makes them diverge in effect becomes primordial here, and there is no longer anything to guarantee that they will agree to meet. That's why the ecological question will now require the properly speculative risk of leaving the terrain of science, where the notion of practice took on a meaning for me, and of adventuring to a place where, it is supposed, only passions, superstitions and illusions, destined to public discord, reign.

TROUBLING THE
PUBLIC ORDER

The Powerless Public

Scientists have learned to scorn the public because it has come to take on the role of a foil in an epoch in which there is no longer any place for Sagredo, the cultivated man of goodwill who unfailingly agreed with Galileo against Simplicio, the representative of traditional obscurantism.[1] At the same time, this public has replaced the people as the referent of politics. Thus one now talks about public space, the space in which one has to debate the stakes for societies that must be democratically decided on. The pulse of "its" opinion is also regularly tested, through inquiries and opinion polls.

According to these inquiries and opinion polls, it seems today that the public trusts the sciences, even if an increasing distrust of the "progress that comes from science" is emerging. One might, as a consequence, fear a future in which the "public representation" of the sciences will no longer fulfill its function, in which the link between the sciences and progress would backfire on the scientists, in which the public would "muddle up" the sciences and their allies. The public authorities are getting worried.

Yet the theme of "good governance" supported by institutions that are supposedly apolitical, because they are unelected, implicitly calls into question a definition of politics judged "outdated." We are no longer on the agora, where ancient Greek citizens publicly debated the future of the city. Those who are entitled to participate in negotiations now are not citizens, apparently a badly constructed fiction, but the *stakeholders* in a situation— each one of whom is defined in the manner of a free entrepreneur whose interests determine how the situation matters to him—

as it is stakeholders who are really interested in it. In the background, the eternal laws of the market loom, judges in the last instance of what will emerge from the negotiations, thereby guaranteeing that the general interest will be served, without anyone having to concern themselves with it.

The instability of this reference to the public, which has no consistency of its own but plays roles that only the reference to progress in general can articulate in a "broadly" plausible manner, is nothing to be surprised about. One might, however, ask oneself what resources we have at our disposal for a problematic staging, for making the unknown that was sketched out in the last chapter vibrate. How, once the "public representation of the sciences" stops keeping it at arm's length, is one to avoid the bad utopia of a "public" endowed with the capacity to object, to hesitate, and to create, which so far I have associated with the strength of a practice? How is one to avoid the *stakeholders* of good governance seizing this unknown with enthusiasm, so as to say "it's us; we are the ones who are entitled to become the actors in your scenario, because the interests (the *stakes*) that we champion, those of our respective businesses, confer on us the capacities that you are searching for!" And how is one to avoid the enthusiastic acceptance of the ecological point of view, which supposes the absence of any arbitration in the name of a transcendent interest, by the promoters of a marketplace that itself has neither faith nor law?

It is not certain that the notion of the public can be "saved," freed from the reference to what must be both kept at arm's length and respected when experts are speaking or from its definition as the object of inquiries and statistical analyses. And it's not enough to affirm that the notion of the public is a problematic notion, because the question then is one of knowing what problem it answers to and if this problem is well-posed. What do the utterances that make of the "public" a subject, the addressee of a message, the bearer of an opinion signify?

Evidently I am not the first to ask myself this question, and I will turn now to the proposition of John Dewey, who was a philosopher but also a sociologist and a political thinker. In 1927, Dewey published *The Public and Its Problems*, which he presented as a response to Walter Lippmann's *The Phantom Public*, published in 1925. Lippmann had shown that the modern citizen postulated by the democratic regime constituted a bad fiction. An indefinitely ready-to-serve being, capable of informing himself about anything and everything, of forming an opinion—preferably an enlightened one—on every question, in such a way that, whatever the problem might be, voting allows the conclusion that the sovereign people have spoken to be drawn. For Lippmann, the shifts of opinion by which the public "makes its presence felt" were rare and sporadic: they always signal a loss of trust in the manner in which professionals manage a question, and once the public receives signals likely to reassure it, to indicate that the professionals have taken things in hand again, it goes back to sleep. The phantom is pacified and no longer troubles the operations of government.

Lippmann's analysis has the aura of a lucid truth, a lucidity that today is more seductive than ever. Thus the GMO scandal can be read in a Lippmannian way: opposition to GMOs translates a loss of trust, and that is what it is a matter of restoring. The response of those who govern has indeed been to invent regulations and procedures that should allow the duly satisfied public to take no interest in the situation, to rediscover its blissful indifference.

It will be noted that the relation between what is called "the public" and the operation of keeping at arm's length that I have associated with the "public representation of the sciences" is being generalized here. There isn't any knowledge, any capacity to object associated with this public, only a sensitivity toward signals that arouse its disquiet or reassure it. But it is, on the contrary, the capacity to object and the link between this capacity and a production of knowledge that Dewey would associate with "his" public.

In *The Public and Its Problems*, Dewey acknowledges his "debt" to Lippmann, but he arrives at "diverging" conclusions.[2] To do this, he begins by changing the temporal horizon, that is to say, by posing the question of relations between the state and the public since the beginnings of the state, thus unlinking the public from the democratic state. Consequently, the mode of existence of the public is no longer that of a discrete or insistent phantom. The public, in Dewey's sense, *emerges*, and its emergence has produced consequences that the state, such as it functions, inherits.

According to Dewey, a public emerges, passes into existence, when the indirect consequences of the activities of a group are perceived as being harmful to the interests of another group, and this latter succeeds in making a "public issue" of what had, until then, been "private," produced by a human association pursuing its own interests. It succeeds in making what didn't count "count," in organizing itself so as to demand the taking into account of consequences that, until then, were accepted as (unhappily) part of the order of things. For Dewey, the state is something that is born and will develop on the basis of such public issues and the correlative perceived necessity of regulating the activity of what can already be called *stakeholders* pursuing their private interests. What necessitates the intervention of a state cannot be deduced from a general definition; it follows from perceived problems, historically produced by the emergence of a public.

So, then, there isn't "one" public facing the state but a history of the emergence of publics, each one of which makes for an event, makes new responsibilities exist for the state. And this is so for better or for worse, as what Dewey describes has no higher justification than the emergence of groups defined by interests the legitimacy of which will be recognized and "represented" by the functioning of the state.

For Dewey, Lippmann's phantom public, which is perfectly incapable of producing a problem, which asks only to trust and to take care of its own affairs, translates not to the (disappointing) truth of democracy to which one must be reconciled but rather to the failure that constitutes what we call democracy. The establishing of what are called democratic regimes coincides with what Dewey calls the "eclipse" of the public. The modern state described by Dewey is not characterized by a knowledge that would be beyond the "simple citizen" but rather by its blindness, the static retaining of rules of functioning that are no longer appropriate to social activities that have dynamics that are always more entangled and uncontrolled. This blindness arises from the rarefaction of the dynamics of emergence of publics, which are also dynamics productive of the mode of intelligibility that defines public matters. The modern "democratic" state perceives itself in a mode that primarily translates to the nonemergence of questions the taking into account of which would have forced it to reinvent itself.

Dewey's diagnostic might seem surprising: Aren't we drowning in information, and isn't the state regularly denounced as producing too many rules and constraints? But what Dewey is signaling with the nonrenewal of perceptions of what is a "public issue" is a disconnection between "information" and "learning." While the indirect consequences of technical and industrial development are tangled in a more and more impenetrable web, the categories of the state remain organized around a division between the public and the private that it attempts to conserve as much as possible.

The powerlessness of the public in Dewey's sense is thus not a "sociological fact" that would have to be accepted, with the mourning of the loss of belief in true democracy this may mean. Nor is the nonemergence of that which, in organizing itself, became capable of forcing "information," what "is well known," to produce consequences, "simply" a political problem. The term *eclipse* can be understood literally: the disappearance of the light that permits visual perception. It really is the source of what renders problems perceptible, in the sense that perceptions produce consequences, that has suffered an eclipse.

A matter such as that of global warming gives Dewey's diagnostic a terrible actuality. We "know," but this knowledge doesn't have the power to produce consequences. And all the explanations that can be given for this powerlessness precisely presuppose the eclipse of the public, the nonemergence of those who would constrain institutions to learn rather than to bury their heads in the sand.

For Dewey, the answer to the question of the "eclipse of the public" is not an "improvement" of democratic life, as if the modern state, which presents itself as democratic, was unfortunately handicapped by the nonemergence of the public. The handicap is anything but contingent. It is the modern state that institutes the fiction that makes political democracy the translation of rights assignable to "bare" individuals, abstracted from any belonging. Correlated with this is the enforcement of the "economic freedom" vindicated by industrial development that claims to be subjected to "natural laws" that politics should not interfere with. In other words, Lippmann's public can only be a "phantom" because the very possibility of speaking of the "public," in the modern sense of the term, is inseparable from its becoming phantomatic, from the "eclipse" of publics whose emergence created public issues, their scope and legibility.

The efficacy proper to Dewey's reading is to break with a point of view that makes of today's democracy an advent on the basis of which general conclusions could be drawn. And that this reading results in an aporia is not, from this point of view, a defect: Dewey doesn't know how the "great community" will come into existence, the community in which "the everexpanding and intricately ramifying consequences of associated activities shall be known in the full sense of that word, so that an organized, articulated Public comes into being."[3] As a philosopher, it's not for Dewey to discern the path, which it would then "suffice" for others to follow; he shares the perplexity of the epoch, and it's with this perplexity that he works, without having the power to transcend it. What he produces is a possibility for resisting the ensemble of discourses that forego this perplexity, which present the situation, in a first approximation, as "normal." The power of the concept of the "public" is thus to make exist, to make vibrate, a "we don't know" that is capable of resisting the "we knows" that our reasoning yields to.

But John Dewey interests me for another reason. What I have just presented is a thesis in political philosophy, but Dewey also wanted to be a practitioner in the social sciences. And the practical proposition that he associated with the question of the emergence of publics seems apt, as we will see, to respond to Georges Devereux's cry[4] calling for scientists who have had their brains returned to them. Dewey championed the social sciences—and here we must understand not just sociology but the ensemble of knowledges that bear on human associations, such as psychology and the sciences of education for example—which might be capable of contributing to the resumption of the dynamics of emergence of those organized and articulated publics that would allow one to speak of democracy.

Dewey's proposition has the great interest of bringing to the fore the question of what would constitute success for such sciences and, in doing so, of suppressing every resemblance between the kind of success that makes experimenters dance and the success he proposes to associate with the social sciences. I have introduced the question of the public on the basis of the ecologico-political problem constituted by the "public representation" of the sciences as a *matter of concern*. But the practitioners who produce the knowledges associated with this *concern* have hitherto been absent. It is to them that we now turn.

The Sociologist-Inquirer

Dewey's proposition offers the great interest of linking the production of knowledge and this very particular "troubling of the public order" that a dynamic of emergence of a public issue constitutes. In so doing, it breaks with what seems to be the doubtful privilege of the social sciences: the sociologist seems to be at home everywhere; his method allows him to approach every situation from its "social" angle, everywhere finding "material" for the application of interpretative or explanatory categories "beyond" the beliefs, convictions, or representations of those whom he addresses.

For Dewey, the "social" sciences definitely did not have to "explain the social" or provide a theory of it or submit it to categories that would finally be scientific. That would have constituted a conceptual error, because the ambition of explaining the social implies, in counterpoint, the staging of a situation imagined as "asocial," in contrast to which it would be a matter of understanding the emergence of associations. But the asocial as primary situation is a poor fiction: association is the mode of existence of everything that exists, of us ourselves, and of everything that we concern ourselves with.[5]

TROUBLING THE PUBLIC ORDER

Dewey suggests that the specialists of the social sciences place themselves in renewed continuity with the experimental sciences by adopting what he calls an "experimental logic." The aim of experimentation here is not to produce a knowledge that transcends concretely particular associations in the direction of a conception of "society" but to learn how a concrete situation is susceptible of changing. It is inseparable from an "inquiry," the primary aim of which would be to participate in the emergence of a group that has become capable of identifying itself and of elucidating—in the sense that an elucidation is a production—its interests in a mode that can possibly constitute them as a "public issue."

For Dewey, the inquiry has to be raised by a difficulty, a trouble, that the inquiry interprets as an obstacle to overcome or a problem to resolve. That is why the inquirer is not, any more so than the experimenter, an objective, neutral, empirical observer but a learner, who is looking for how a situation "holds together" and how it can be transformed, who creates knowledges that do not designate an object or a fact but a learning trajectory nourished by her own attempts at modification. What is more, the inquirer will be an experimenter in her own manner, because the success of the inquiry implies the creation of a successful relationship between the inquirer and the inquired. The difficulty that initially engages the inquirer, which "troubled" her, can, in effect, be overcome only with those who have raised this trouble. And her success is translated by the production of tools conducive to allowing the group to articulate what it lives and undergoes. The goal of experimentation is therefore not one of allowing the inquirer to discover the reasons for the situation that has troubled her but of participating in the production of these reasons: "at the end of a series of trial and error tests, the inquirer and the inquired agree on a point on which the cognitive experience of the former and the life experience of the latter enter into relation, are felt and redefined in relation to one another."[6]

Unlike a convention in the usual sense, which articulates already identifiable interests, the point on which the inquirers and the inquired "agree" succeeds in making a relation, in bringing terms that did not preexist this agreement, that arose from a double process of the production of existence into relationship. I cannot imagine the Deweyan inquirer dancing on her terrain, but I can imagine her uneasy joy when those with whom she has worked seize hold of some of the words and knowledge that she has learned with them and make them their own, producing the "reasons" for their situation. It means that the group has produced a new regime of existence, that the words and knowledges associated with what it has undergone have now got the power of raising a "public issue." Sociology here is not "subjected to politics" in the sense of an already existing politics, but rather it enters into a symbiotic relationship with a politics of emergence. To learn something new, it needs people who will put the relevance of what it learns at risk, that is to say, whose emergence will make the difference between success and failure. But this apprenticeship is not an end in itself; it is correlated with the political event constituted by the emergence of the group that has become capable of intervening, of arguing, and of struggling.

Let's not fool ourselves. At the moment when Dewey made this interesting proposition, he was not a marginal thinker. His renown was at its peak. And yet he was witness, in the most radical powerlessness, to the "professionalization" of a sociology that was serious and methodical at last, supported by the "facts," as is appropriate for any real science.[7]

It is certainly possible here to refer Dewey's failure to "big" reasons that explain why things couldn't have been any different. Not only did his proposition collide head-on with the ideal of a politically neutral science, but it also deprived sociologists of their abundant "matters of fact," the subject of the innumerable writings that the profession is nourished on and that make of the sociologist (or the psychologist) the person whom one asks to explain, in neutral terms (by referring to respectable "matters of fact"), the meaning of anything that "troubles" the public order. The defect of these reasons is not that they are false but that they do nothing to advance the question that interests me here, that of the practice that Dewey proposes.

In order to learn, then, I will play devil's advocate: I will behave "as if" sociologists had had interesting reasons for having rejected Dewey's proposition. This "as if" arises from a form of thought experiment, the stake of which is an attempt to bring Dewey's "experimental logic" into relationship with the notion of "practice" that guides me.

Dewey did not think in terms of practices, that is to say, he didn't ask himself what had the power to bring practitioners together, to oblige them to think and make them hesitate. Quite the contrary, he attempted to situate the sociologist inquirer that he was proposing in a relationship of maximum continuity with experimental, laboratory sciences, and he put them all together in the lineage of the experience of living beings. The history of living beings would itself be a history of trial-and-error tests, with the emergence of successful connections that codefine a living being and its milieu in terms of a transaction. The same logic—experimental logic—would prevail in all cases, as creative response to an obstacle, as reconstruction of experience in a new mode, implying a new "transaction" between the active being and the world that affects it.

From my point of view, this continuity is not "false," but it has the defect of generality: every case becomes an example illustrating this generality . . . including the activity of sociologists who, wanting to be "scientific," refused Dewey's proposition. In effect, if for them the obstacle to overcome is primarily the accusation according to which, far from doing science, they are engaging in politics, not restricting themselves to describing states of affairs but contributing to the activation of restless minorities, in the name of what are they to be denounced? The choice of thematizing their activity in a mode that makes professional neutrality matter *is no more criticizable* than that which places the action of the spider under the imperative "killing—eating."[8]

The sociologist proposed by Dewey certainly might be "troubled" by this choice. But in this case, the trouble does not communicate with the possibility of knowledge production. Inquiry, for Dewey, effectively asks that those it concerns share with the inquirer the sentiment that, in one manner or another, "it has to change," although, for one reason or another, they are incapable of defining collectively, by their own means, how to construct the reasons for this sentiment in a mode that allows it to pass into public existence. If the refusal of Dewey's hypothesis is inseparable from the professionalization of sociologists, from their will to conform to the common claim of scientists, the claim to autonomy, *in the terms of experimental logic*, that refusal is not a matter of inquiry.

It gets worse, and it is here that the question starts to become interesting. Because those who would want to defend Dewey's proposition against the choice sociologists made to professionalize may well foreground an ethico-political norm that would transcend experimental logic—namely, being in the service of a living democracy. Now, to accept such a norm is synonymous with putting the social identity of sociologists as professionals in danger: Why would they subject themselves to it, when physicists or chemists first serve the interest of their science? This, it seems to me, is the blind spot of Dewey's proposition. The strong bond that he establishes between sociology and democracy creates an unnegotiated break with an aspect of the model of the experimental sciences that he claims to prolong: "experimental logic" doesn't have any privileged link with the general interest.

When, for example, physicists hesitate with regard to solar neutrinos, for sure they are not neutral; they are engaged, in the sense that they refuse to avoid the difficulty by means of a simple convention among physicists. But they owe it to themselves to maintain a disciplined neutrality with regard to the eventual outcome of their hesitations: each physicist can have his or her favorite hypothesis, but what counts for all of them is that the hypothesis that will impose itself be authorized by successful experimentation. The repercussions of, for example, neutrinos gaining a mode of existence endowing them with a mass, possibly acquiring new importance, making a difference for new actors who will say "thanks to neutrinos we can now . . ." are contingent, in the sense that the protagonists of these consequences—funding bodies, instrument manufacturers, promoters of a diverse variety of programs—may well prepare themselves, speculate, provide resources, but they are supposed to be held in abeyance for as long the physicists hesitate.[9]

The Deweyan inquirer, by contrast, cannot have such neutrality. The stake of her work is not simply the transformation of a situation that allows for inquirers and inquired to be brought into relation, indeed to become coinquirers. It has to be a matter of a transformation that produces new capacities for situating oneself on the part of those who are "in" the situation, a transformation the value of which derives from its political consequences. Learning how to inquire about what may succeed in transforming a group into a sect or a blindly fanatical association isn't part of Dewey's program. In other words, what engages the Deweyan sociologist is instead the success of an "engineer" for whom each local success punctuates a trajectory the primordial purpose of which is not knowledge but the "improvement" of a mode of functioning, here a sociopolitical functioning.

For Dewey, sociology ought to work on creating knowledges that affirm and activate the possibility of democracy, because this latter doesn't constitute only one mode of organization of public order among others. As a philosopher, Dewey maintained that it was a matter of the political regime most propitious to the accomplishment of being human as such, the mode of existence of humans being one of communication and of shared experiences. Taking up Dewey's problem again is not to ratify the simple idea according to which this philosophical conviction is sufficient to explain his failure, a "true science" having to be autonomous in relation to philosophy. But it is to affirm that it's not for philosophy to short-circuit the question that Dewey, it seems to me, did not broach, that "experimental logic" did not engage him in broaching: the question that I associated with scientific practices, which Latour describes in terms of links and knots. Dewey's strength was to think the knots, the production of the relation between the practitioner and what she is dealing with. His weakness, perhaps, was not to have thought what links sociologists themselves as practitioners.

Humans as Such?

It is thus as much his strength as his weakness that has made me take Dewey as an "example." Dewey's proposition has retained all its actuality. One may think that if he were still alive, he would have praised the manner in which learning trajectories, trajectories of knowledge production have been produced around the question of GMOs, catching industrialists as much as the state off guard, or in which groups of drug users have arisen, capable of producing a knowledge that made specialists who defined them as "criminal or ill" stammer. He would also, no doubt, have seen in the question of global warming the model stake that makes his bet against Lippman, that the "powerless modern public" is not the last word, matter: if Lippmann is right, we will undergo global warming in the worst possible of ways. But what makes for Dewey's strength is not just the political relevance of his analyses; it's also the passage that he accomplishes from the question of knowledge to the question of a production of existence.

I have already insisted a great deal on successful experimentation as the creation of a relationship. It really is a production of existence, the production of terms as henceforth "related." But in this case, the relation can easily be reduced to a condition for a knowledge that is finally authorized by the facts that have arisen from it. With Dewey, the production of existence is not just first, but the new knowledge, which is inseparable from it, cannot reduce this production to the status of a condition, which can be forgotten once this knowledge has been acquired. The emergence of a new public, in Dewey's sense, can be explained from the moment when what the group protests against is recognized as an effectively intolerable nuisance, which it is normal to organize against. But "to be explained by" dissimulates the event, the becoming of the group that constituted itself around what will subsequently be recognized as a legitimate reason for a modification of the definition of public order. The nuisance around which the group constituted itself cannot be identified with the nuisance such as it may eventually be characterized in terms of public perception and then regulated in terms of the state's perceptual apparatus. Characterization and regulation are part of the manner in which the event "is explained," that is to say, in which, as part of its consequences, there is a production of regulatory words for speaking at one and the same time of what has been perceived as unbearable and of what is organized around this perception. In other words, what will subsequently be described as justifying the event—the production of a new public order—is part of the consequences of the coming into existence of a public grappling with the unbearable situation that made it emerge.

However, the production of existence that marks the passage from a "private" suffering to a "public" issue is a very particular case. The weakness of Dewey's proposition is perhaps that it doesn't offer the means for thinking this particularity, which it drowns in the generality of the continuum it inherits and which it means to renew. In biology, this continuum is usually placed under the sign of adaptation, and in human histories, it results in putting "man" and his capacity for inventing solutions to the problems he is posed at center stage. In Dewey's continuum, trouble and transaction replace adaptation, but they also result in the relative abstraction of a "troubled human," carrying out a reconstruction of his experience, valid as much for a child at school, an emerging public, and a scientific practitioner.

And it is indeed because he claims to prolong a general logic that Deweyan sociologists cannot protect themselves from the accusation that confuses them with "social reformers," working for the common good, human emancipation, the progressive amelioration of the definition of the public order. On the one hand, the "trouble" that animates them—the existence of a weak group, which undergoes something without being able to protest in a way liable to make a "public issue" arise—makes them the worthy representatives of a concern that all humans as such should share. On the other hand, one cannot see in what way this trouble is susceptible of making them *hesitate with their colleagues*, in a mode that leaves the question of public order in suspense. In effect, nothing obliges them or their colleagues, other than the cause of humanity and its progress, and so nothing protects them against a general goodwill that allows what they do to be assimilated to what the world as a whole ought to do.

As usual, the case of practitioners who happily accepted the role of the goose with the golden eggs is significant in this regard. To be sure they agreed with satisfaction that their research be defined as the very motor of human progress. And in their own way they took the consequences of this definition seriously. The ease with which scientists have, in this framework and with perfectly clear consciences, "labeled" what they do—adopted a presentational rhetoric destined to attract research funding associated with a problem of general interest—does not have a lot to do with the blessed indifference of the goose. It is probably not, primarily, a proof of their cynicism, but it does translate the fact that what links them, what obliges them, is elsewhere.

The role that Dewey proposed to confer on "inquiring sociologists" in the emergence of new "public matters" links sociology with a proposition that bears on "the human as such," whose great achievement corresponds to a democratic regime. The question that poses itself is this: Is taking charge of the interests of the human as such suitable for a practice, whatever it may be? This question is the correlate of the ecological point of view that I am trying to adopt, a point of view that foregrounds heterogeneity and poses the question of relations as relations between the heterogeneous. We don't know what belonging to one's practice makes a practitioner capable of, but from the ecological point of view, this question can never be short-circuited by an appeal to values that would remind us all that before being practitioners, we are humans.

How can this ecological point of view renew Dewey's proposition regarding the practice of "inquirer-sociologists," this proposition whose major interest is, as has been seen, that it designates the situation in which the production of knowledge and the production of existence are inseparable? I will refer here to the conclusion that Bruno Latour has drawn with regard to the anger of scientists when their facts, theories, and controversies have had a mode of "social explanation" applied to them, which was no more insulting than that which sociologists have used in many other domains, from art to medicine, passing via religion or accountancy.[10] Latour speaks of a *felix culpa* or a "fortunate fall from grace" in the manner of Adam's sin, which for certain theologians was providential because it opened the history of redemption. The question, then, is not one of knowing why scientists judged that they were right to protest but much rather why others have not. And the response is, of course, that this time the sociologists aimed "higher" than themselves, at the representatives of rationality, objectivity, the proof, and "lower," at those who, in any case, knew themselves to be deficient in relationship to this standard and thus did not feel equipped to contest the (scientific, objective, etc.) approach that judged them. In other words, sociologists benefited from the scientific authority that they intended to deconstruct. For Latour, one must agree to talk of a failure of the "social explanation of whatever" so as to be able to admit that where and when sociologists engaged in it with impunity, this was a "fault," a happy fault, if it opens the path to redemption.

The path that Latour calls his sociologist colleagues toward is a path on which *we would learn to treat humans as well as we are capable of treating nonhumans.* In repeating this wish, I am choosing to expose myself—that is to say, to provoke the ire of all those who think of themselves as "humans" and who have already written lengthy denunciations against anyone who would dare to treat humans as if they were pebbles, solemnly reminding those who would forget the radical ontological rupture that passes between pebbles and humans. Now, it doubtless hasn't escaped Latour's attention that it is humans who speak of pebbles and not pebbles of humans and that he himself addresses readers and not pebbles. What apparently escapes the alarmed protestors is that Latour wrote "as well as" and not "like," that he set a challenge for his sociologist colleagues but did not propose to assimilate humans and nonhumans, be they rats, baboons, or pebbles.

The category of nonhuman is powerless in and of itself to authorize any kind of assimilation. How is one to gather together, under the same category, neutrinos, genes, bacteria, bonobo monkeys, tornadoes, oceans, motorbikes, pigs, and mosquitoes? And how is one to dare, apropos of this heteroclite ensemble, to propose that humans should be treated as well as we treat all members of this ensemble? Think of livestock, subject to more or less the same definition as exploited workers, except that mistreatment is even more unrestrained in their case. Or of the female babies of rhesus monkeys that, not so long ago, the experimental psychologist Harry Harlow took as a matter for testing and subjected, in the name of scientific proof, to the worst kind of "mistreatment": he wanted to establish a cause-effect link with a future in which, once they were adults, they would mistreat their own offspring.[11] Also, to treat "as well" certainly doesn't mean "treat in the same manner": one doesn't treat a virus, a molecule, a monkey, or a motorbike in the same manner. But it could mean constituting what one addresses as that from which it is a matter of learning how to "treat *it* well," that is to say, when it is a matter of scientific practices, how to treat it in a mode that permits the learning of relevant novelty.

One could, then, take up again Dewey's characterization of the undertaking of his inquirers without having to associate them with too-rapid a convergence between scientific practice and the progress of humanity. The definition of success for the Deweyan inquirer—to make "treating well" coincide with participating in the production of a group capable of transforming what it undergoes into a "public issue"—is relevant but not generalizable. More precisely, it is what makes it pertinent that should be generalized. It's because the emerging group is capable of feeling insulted by certain manners of characterizing the cause that it defends that the sociologist-inquirer can learn how to "treat it well," that is to say, how to enter with it into a relation that conjugates production of knowledge and production of existence. Could hesitation with regard to what "treating well" signifies in each case, to what it takes to address groups well, who require not that they be addressed as "humans" but on the basis of what brings them together, bring sociologists together, separating them also from any consensual reason? Indeed it would be a matter for them of learning to treat equally "well" practices celebrated as serving reason and progress, others that are reputedly irrational, such as the practices of pilgrimages to the Virgin Mary, or still others, activist practices—when the suffering inflicted on animals or on humans becomes what obliges thinking and acting.

There might be a point of agreement between the position being defended here and that of Jacques Lacan, when he remarked that "the human sciences no more exist than do small savings." Every penny counts: no hesitation is pointless. References to the intentionality, the freedom, the symbolic thinking that are the privileges of the human may well be put in danger by Latour's proposition, but are they not also what allows humans and their associations to be subjected to all-terrain generalities? And indeed Latour united against him not "defenders of humanity" but rather his own colleagues, those whose job it is to deal with and to "treat" humans? Those whose categories are founded on the "privileges" attributed to humans?

But if the category of the human is a shortcut suiting those who, for one reason or another, yield to the all-terrain generalities through which those humans are characterized, there is another little shortcut brought into question. This shortcut is the reference to a "common humanity" that might allow the conflicts that oppose humans to each other to be situated from the point of view of a consensual reason to be established, a reason that all, because they are human, are supposed to adhere to. And here, it is no longer a matter of the practices of sociologists but rather of the question posed by the coexistence of practices that, insofar as they refuse to submit to "consensual reason," can be associated with disturbances to "public order."

Thinking Through Causes

Dewey had underlined the difference between associations in the general sense and the emerging public. But from the practical point of view that I am going to explore from now on, the difference passes instead between association in general and those particular associations, which I will now call "practices," including scientific practices, that can only be "described well" by taking into account what obligates those that they bring together—that which can be called a "cause."

The notion of "cause," such as I am introducing it here, is familiar to us when it is a matter of a cause of a political, moral, or religious type. But one can equally talk of a cause when it is a matter of a nuisance being recognized as a "public issue." It is because an emerging public has a cause that the Deweyan sociologist can enter with it into a relationship that will allow him to learn: this group is capable of hesitating, of making the difference between a successful and a failed relation. And what obligates experimenters is, in this sense, equally a cause—that which sociologists experienced when they found themselves accused of "treating badly"—of insulting—the practices of their experimenter colleagues, while they (the sociologists) were "just" using an approach that intended to describe scientists in the same manner as they would describe any other group. Even Dewey's notion of transaction, which purported not to be reductive, would have provoked the same accusation, if transactions had been defined as conventions, having the aim of finding an arrangement between diverging interests.

Sciences are certainly not practices "like" the others. No practice is like any other. But it's a matter now of thinking them as practices "among others," that is to say, not only of refusing, for other practices, too, the general categories, which insult scientists, but also of accentuating the way they diverge, each for its own sake.

I will take as my first example groups of (unrepentant) drug users. To be sure, when such groups have undertaken to make the consequences that outlawing drug use had on them into a public issue, they wanted to be recognized as "citizens like everyone else." But in another mode, certain of them did not take themselves to be "like everyone else" at all, because they were also united by another question, that of the relations to be cultivated with "their" drug. And the "cause" that then unites them has nothing to do with what unites scientific practitioners. There is, perhaps, a matter of a common knowledge to produce with regard to drugs, including the molecules that are produced in laboratories, but the users are not themselves "colleagues" raising objections and seeking to assure themselves that it is indeed with regard to the drug as such that they are learning. It's a "user culture" that they are seeking to construct, and they have to do so in a landscape that has been ravaged, in which most of the cultures that associated the consumption of what we call drugs with practices that confer on them significations that are irreducible to what we call "psychotropic molecules" have been destroyed. It is a matter for them of learning to deal with something that they know very well can be identified in terms of chemical compounds, while also knowing that this definition has no relevance for them. If we have to speak of a "transaction" here, it is a strange transaction indeed, one that undoes the general categories assembled around "human" and "molecule," because the relation to create will be, inseparably, the bringing into existence of what we call a drug as a powerful being, as a "cause" obligating to learn appropriate "usages."

Can one talk of a "transaction" when it is a matter of ethologists of whom one will say, in accordance with Devereux's wish, that they have succeeded in keeping hold of their brains: they are learning to observe, even to interact with, animals in a mode that allows these animals to testify to what they are capable of. Perhaps one can, but here, too, what comes to the fore defies any generality, because the ethologist must try to confer on her bonobos, crows, wolves, sheep the power to make her hesitate with regard to the way of "best describing" them. The practitioner herself cannot be "described best" without making a "cause" intervene, that which makes her hesitate in a mode that she doesn't submit to, that doesn't "trouble" her, but that makes her a scientist, able to provide relevant knowledge to her colleagues. The knowledge that she will provide for her colleagues is not "objective" in the sense that those animals she has learned to describe would be required to undergo counterproofs that aim to ensure that her description is authorized by the animals independently of her relation with them. Her colleagues, if they have not been deprived of their brains, as Devereux would say, will know that what she reports is inseparable from the manner in which she has herself been transformed by what she has had to learn in order to enter into relationship with those she describes, and it is as such that this report will interest her colleagues. If they are able to affirm that what binds them together is the learning of ways to relate to animals, not the answers they obtain from the animals, those ethologists will be able to affirm that their practice is not a scientific practice "like the others" but *among others.*

CHAPTER FIVE

It is rather different in the case of experiments, this time undertaken in the laboratory, that can certainly be placed under the sign of transaction (the animals are rewarded for their performances), but where the vocation of the transaction is not to "describe well." It is rather to intervene in an active manner so as to "raise," or "uplift," the animal, to bring it to enter into hitherto humanlike relations with humans. In this case, experimenters endeavor to learn which situations would allow their animal "pupils" to explore new "possibilities of existence." At the limit, it's a matter of producing veritable "chimeras" such as great apes that, it is said, learn to "talk." Orangutans, chimpanzees, gorillas, or bonobos here become a cause, not for hesitation over the best manner of describing them but for speculation bearing on what they might become capable of.

Employing the term *chimera* doesn't signify an opposition between the bonobo "in itself" and the bonobo "for us," the product of our speculations. In all cases we are dealing with a bonobo "in our midst," in the midst of a dense and conflictual weft of manners of mattering for us; we are dealing with a bonobo, whose "consistency" derives from its manner of making hold together the divergent bringings into relationship that it lends itself to. Correlatively, "lending itself" must be understood in the strongest sense of the term, not in the sense of a receptacle that is indifferent to our projections, and not in the sense of the "reliable witness," synonymous with successful experimentation, either. Thus whatever the bonobos do, they are not "reliable witnesses" to a "potential linguistic capacity" attributable to bonobos "in general." The statement "he knows how to speak" (or sign) is too heavy, laden with too many stakes, too indeterminate to be defined as the affair of a group of competent colleagues. And the bonobo that is said to "be able to talk" is always an individual inseparable from the highly singular milieu in which it has become a "speaker." Yet the "success" of these specialists really is the coming into existence of a new being, coproduced with humans who have, for their part, had to learn to discover what "raising an ape" demands.

The purpose of a success like this is not "describing bonobos, or other 'gifted' animals, well." To the extent that their "breeders" or "teachers" have understood that they are not "revealing" a potentiality of their pupils but are participating in the coming into existence of something new, the controversy over facts and artifacts has lost the power to make them hesitate. The question, "What can a bonobo do?" no longer concerns "the" bonobo, such as its interactions with humans allow it to be approached, but *this* bonobo, which answers to its name, which a life "with" humans has brought to existence.

It's interesting to note that a problem (which also marks the human sciences) is posed for those teachers—namely, that of the responsibility of researchers with regard to the becoming that they are engaging their subjects in. Animals that learn in a lab are certainly indifferent to the stakes that make their "teachers" speculate. But they are not in the least indifferent to the care, the attention, the multiple and varied solicitations without which they wouldn't be learning. The bringing into relation that allows them to do what specialists call "speaking" involves care and attention, even intimacy, which becomes constitutive of their lives as such. They depend on a milieu where they are spoken to and with, addressed as speaking persons. What then to do with these chimeras when they are aging or, in the case of a research team, being dismantled? Can they be doomed to the misery of survival in a milieu in which they would be nothing more than inadequate monsters? The "teachers" then get funds for a "cause," which is not that of the "advance of science" but of creating "retirement homes" for those who, they feel, they have to answer for.

The example of talking apes is suitable for raising many hesitations. Do we have the right to create such chimeras? Aren't we abusing our powers? The contrast with the economic and rational justifications associated with the transformation of farm animals into meat on legs, which "zoo-technicians" busy themselves with, is striking. But so are the ready-made justifications, in the name of the conquest of knowledge, that intervene, as long as the justification can be assimilated to the production of a reliable witness and not that of a chimera. In the case of psychotropic drug users, the transaction, or bringing into relation, is explicitly a production of existence. And it's the perplexity that this explicit production provokes, in controversies, derision, denunciation, that matters to me. That's what it is a matter of cultivating.

If it's a matter of cultivating this perplexity and not of crushing it with abstractions (does man have the right to . . . ?) or order-words (the irreversibility of advances in Science or of scientific innovations), it is important that the cause that engages the practices that provoke this perplexity be recognized as "positively nonscientific," with reference to a scientific type of knowledge. This kind of reference, in fact, weakens them. Drug users will be ridiculed if they seem to forget that a drug is nothing more than a molecule (plus human subjectivity); and the specialists of talking apes will be denounced because they do not reveal a preexisting potentiality in their animals but instead fabricate chimeras. Let's not mention teachers, to whom pedagogues and epistemologists are incessantly explaining what they unknowingly do and what they should do, now that they know. Cultivating perplexity demands that what causes perplexity be "described well," in its divergence, as it happens. And this divergence, in return, illuminates what every practice of bringing into relation, scientific or not, implies: the inseparability between the relation and the bringing into existence of what is brought into relation.

Exhibiting One's Divergence

At the start of this chapter, I turned to the manner in which John Dewey addressed the question of the "public" and separated it from what the "public representation of the sciences" requires and presupposes, a public that is a bit flighty or a bit of a phantom (to borrow Lippman's expression) and that it is a matter of pacifying, comforting in its role of passive beneficiary, and that—above all—mustn't meddle with something that nevertheless concerns it. Dewey's emerging publics, which trouble the definition of the public order, demanding that what matters to them be taken into account, contribute to making the unknown of what we call "the public" resonate. This unknown is nothing other than what the very idea of a political democracy demands, if it is to be defined in contrast with a sophisticated art of leading a herd.

What Dewey called the eclipse of the public signals its reduction to a herd that has to be guided, reassured, edified. To affirm that this is an unavoidable reality may seem "realistic," but it's a vicious realism, as it is not possible to talk of the public independently of the manner in which it is treated. If what we call "public" is contemporaneous with the division between what is of the order of "objective" knowledge and the "rest"—the beliefs, fears, values, or particular interests, which "distort," as French researchers wrote in 2004[12]—we can only speculate on the question of how things would be if this a priori division was proscribed, if the complete unfolding of every "matter of concern," in Latour's words, was defined as calling for the gathering together of everyone it concerns, of everyone who has a *cause* to be concerned. Maybe we can imagine that what we call "the public," those who are generally concerned but who, in *this* case, have no cause to defend, could get a positive, even crucial, role. "Making public" the unfolding of matters of concern is indeed crucial to opposing the dismembering of an issue. Speaking in the presence of an attentive, interested public is what stakeholders fear; it can force each concerned party to make its "cause" exist without negating those defended by other parties, without mobilizing the consensual reasons that no party can represent.[13]

In ecological terms, abandoning the idea of consensual reason might be phrased as abandoning the ecological model that knows only predators and prey as a way of thinking of public order—the predators being those armed with the right to silence, to keep at a distance, to disqualify others. It then becomes a matter of thinking what, from now on, I will call an ecology of practices, or an ecology of causes, in the sense that these causes obligate divergent modes of thinking, feeling, and acting. Such a point of view has nothing "reassuring" about it, in the sense that one cannot expect from it a new mode of assigning legitimacy. It is not a guarantee of a solution: don't ask me how I intend to found the democratic right to struggle against the enemies of democracy or if I really believe that such and such a group of fanatics will agree to a peaceful coexistence with those whose activities it judges as anathema. It is not, then, a matter of providing a general answer where hesitation must prevail, where decisions that commit have to be risked. Rather, it is—and this is why the term *ecology* is adequate—much more a matter of thinking a situation in which the decision concerning "what we are going to do" is never redoubled by an "as is our right," demanding the adherence of others. No "small savings." Every cause that obligates one to hesitate may matter.

It's to philosophy, since its Platonic origins, that we owe the thesis according to which there must be criteria that transcend conflict—even if this transcendence is disguised as immanence. Such is the case with Dewey's proposition, according to which philosophy has to abandon "its rather sterile monopoly of a commerce with an ultimate and absolute reality" but would find compensation for doing so "in enlightening the moral forces which move mankind and in contributing to the aspirations of men to attain a more ordered and intelligent happiness."[14] It is thus to be expected that many contemporary philosophers will protest when faced with what a reference to ecology proposes. They will see in it that which it is precisely their responsibility to think against: without any beyond, without anything that can justify our conviction that force cannot be the final word, the wolf will eat the lamb, the strong destroy the weak. The vocation of philosophy, then, is to be the incarnation of the difference between a nature that is "red in tooth and claw" and that which would make of us humans, refusing "nature," the last word. It would be for philosophy to be the spokesperson for that which transcends our conflicts, for that which would make humans come together as humans, beyond the divisions that consign them to war.

With the question of an ecology of practices, I am attempting to inherit what has happened with the role effectively played by philosophers who have sought to ensure a "public peace," a public order in which the sciences find their place. And notably by those philosophers who have participated in the construction of a "public representation" that characterizes the successes of these sciences in a "common" language, in terms of problems that are those of everyone and that these sciences would merely have succeeded in formulating in a manner that is rational and decidable at last. In doing this, philosophers have ratified, even collaborated in, the operation that transforms the strength of these scientific practices into authority and that makes of them the "brains of humanity," speaking the only language able to bring humans into agreement. Even if this means that other philosophers, in the name of other requirements, define this brain—a brain that calculates and is deaf to everything that escapes calculation—as that which must be forced to yield, so as to make a properly human space exist, one in which it will be a question of something that alone ought truly to unite humans: what it is that makes them human beings.[15] An excessive privilege for philosophy, but indignity, too, because it forces philosophy to present itself in a mode in which it becomes an authority, a master of thought, benefiting from a direct relationship to the question of the human "as such."

Evidently I am not speaking here of "all philosophers," and one might even object that I am speaking only of pseudophilosophers. I would doubtless be more or less in agreement, but the question that matters to me here is that of the predictable mode of denouncing the ecology of practices, mobilizing the bogeyman that, since Plato, has played with our fears—irreparable division and the rule of the strongest would be the only horizon if one cannot call on an arbiter with the right to demand that everyone comply. The ecology of practices is not the recognition of a factual state of affairs but a perspective that demands and engages. It engages me and demands not that I present myself in the name of "grand philosophy" but grappling with my philosopher colleagues, for whom philosophy has a role that one might call "pastoral": that of reminding the human race of its vocation, offering it reasons that are worthy of being common to humanity.

From the point of view that I am defending, it's not enough to renounce reasons in common or any position of transcendence. It is not enough for practitioners to say, modestly, "We are just doing our thing" and we don't demand that you take an interest. It's a matter, for each practice, of creating its own manner of actively, publicly, refusing its assimilation to the putting to work of consensual means and reasons. If an ecology of practices is to resist the triviality of an ecology of predators (whoever or whatever speaks in the name of a transcendence) and prey (whoever or whatever is disqualified in the name of this transcendence), it demands that every practice exhibit its divergence, present itself on the basis of what makes it diverge.

This doesn't offer any guarantee. What makes for divergence can lead to conflict. And the engagement that corresponds to this demand accentuates still further the refusal of any conniving with the interests of public order. No practice that accepts that it belongs to an ecology of practices should claim a privileged relationship with these interests or with those of humanity.

This doesn't signify that every practice must present itself as hostile to public order, because that would surreptitiously reintroduce a form of transcendence. It's simply a matter of recognizing that no practitioner—as a practitioner—speaks a transparent public language, one equally accessible to all, that calls on all its speakers to recognize each other as free and rational. To every practice there corresponds instead what can be called an "idiom" in the Greek sense, wherein "idioms" rhymes with the language of "idiots," of those who do not benefit from a language that flaunts its claims to public intelligibility. But if the language of a practice is always an idiom, then the practitioner is always an "idiot," in Deleuze's sense: someone who diverges from the "common," who remains a stranger to what makes for consensus, not because he would have another position to propose, able to generate unanimity, but because *there is something more important.*

For a practitioner then, exhibiting one's divergence doesn't in the least mean presenting oneself in an epic, or even an "aristocratic," mode, as someone who escapes from the common. To be sure, the idiot in Deleuze's sense does not enter into the network of oppositions through which a public, whose powerlessness is a problem, and active protagonists, those whose interests define if and how a situation matters, are put on stage. The idiot's refrain, "There is something more important," does not make of him a judge. Thus, in the history of solar neutrinos, there was something more important for the physicists than "human protagonists" coming to an agreement. But if those physicists felt *obligated* to suspend any conclusion that doesn't make them judges of different situations—those where, for example, protagonists can be delighted by an intelligent convention agreed between them, creating an acceptable articulation of heterogeneous interests.

The idiot, in Deleuze's sense, slows down, whereas others, mobilized by an urgency that can be expressed in consensual terms, speed up. But the idiot doesn't slow down in the name of a transcendence that would allow the disqualification of common humanity or in the name of interests—his own—that he seeks to make prevail. If he is "uncommon," the common of which it is a question is the consensual common required by public order. Yet what matters to the idiot and makes him diverge doesn't enclose him: it's not a matter of the tragic figure of incommunicability, a figure that in fact belongs to public order. The idiot can "pass on" what matters to him, make those who rush around imagine and hesitate, and that is why he really does constitute a "trouble" for public order. He can trouble those who allow themselves to be defined in terms of interests—that is, those who identify themselves in the mode that a public language asks: conflicts of interests are needed in order to legitimate the formulation of priorities and the consensual reasons that are supposed to order them.

To conceive the ecology of practices on the basis of this bet—one's practice makes of the practitioner an "idiot"—belongs to speculation regarding the possible. From the "factual" point of view, certain practitioners obviously behave not as idiots but as *stakeholders*, even as artists in the manipulation of so-called public interests, which they translate in a mode that permits them to present their practices as those that these interests require. As we have seen, the regime of knowledge that is in the process of being dismantled supposes, requires, and gives its blessing to such manipulation as the condition for innovation, so that the eggs laid by the scientific goose are golden. The bet taken by the ecology of practices, then, is not authorized by a state of affairs—"practitioners *are* idiots"—but speculates on the possibility that one can address practitioners as those for whom there is something more important, without insulting them. And that *all* practitioners can be addressed in this way, independently of the role (whether predator or prey) that divides them.

But this bet has a consequence that matters. Not only might practices, all practices, be said to "trouble public order," since they make idiots and idioms exist, but this public order itself, with the consensual interests that everyone must kneel down before, might be described as what commands practitioners to "choose sides," as predators or prey, as being capable of presenting themselves in the mode that public space requires, or not. If one recalls that public order was not born from the general (good) will but rather from campaigns to "pacify," to eradicate, which occurred a long time ago in our countries and more recently in those that were colonized, one might suggest that every practice rather *think of itself and present itself as "surviving,"* in an environment that is fundamentally hostile to it. Scientists are not wrong to try to protect themselves.

To adopt the point of view of an ecology of practices is thus not to denounce the lies of practices that present themselves in terms of consensual reasons but to choose to hear these reasons as survival strategies. However, what this point of view is committed to, exhibiting divergence, is not synonymous with engaging in an immediately political struggle—practitioners of the world unite! This order-word does not suit idiots. Certainly that is the weakness of idiots but also their strength, because if they refuse to be enrolled, they are, by contrast, capable of disarticulating what enslaves political practices, the distinction between facts and values, and what poisons them, the contempt for a public that continues to confuse facts and values. The question, then, is not, "How are these practitioner idiots to be enlisted?" Rather, the question is how to conceptualize a political struggle that takes actively into account what they could be capable of. One thing seems clear to me: an "inclusion" of this sort will be impossible, as long as what predominates is the commonplace according to which if there isn't a way of arbitrating between diverging manners of defining what matters, the war of all against all will prevail. For idiots are those for whom there will always be something more important than such a reason.

The point of view of an ecology of practices doesn't offer any guarantee of peace. On the contrary, it creates an unknown. We know very well what *stakeholders* are capable of, dismembering a situation according to the categories of interests by which they are defined. We don't know what "idiots" might be capable of, presenting themselves as obliged by causes that are not common, that are not addressed to the human as such. This is the unknown to which the proposition of an ecology of practices attempts to give the power to think and feel differently what we had learned to judge as "normal."

INTERMEZZO

THE CREATION OF CONCEPTS

———

No ecological regime, whatever it might be, is able to define its protagonists, each one of which in fact now exists only because it has been able to survive both in spite of and thanks to others, at the same time. Thinking an ecology of practices does not, therefore, permit me to define anything at all, but it does demand that I present myself in a mode that is coherent with what I am proposing. As I am a philosopher, I have to present myself in a mode that displays its divergence. That is why it is necessary for me now to pass from the "notion" of practice to the question of what this notion commits me to as a practitioner of philosophy.

So far I have spoken of the "notion" of practice. This signifies that I haven't yet taken the type of risk that I associate with philosophical engagement. I have tried to "describe well"—that is, without insulting those whom I did not want to insult—a situation that commits and divides, a situation that cannot be described in a neutral manner because the very words whose articulation it imposes do not describe but "serve notice" of the stakes, mobilize positions. This is what was seen with the "science wars," a war of position conferring contradictory stakes on the words *objectivity*, *neutrality*, *autonomy*, and so forth.

Usually the term *notion* designates an intuitive, synthetic, rather imprecise knowledge that one has of a thing, a rather pejorative definition signaling a lack of certified expertise. However, the word could be saved if it is associated with the manner in which the thing in question "matters," the manner in which one will address it. "Not having the slightest notion of . . ." doesn't signify a lack of knowledge but rather the incapacity to situate oneself in relationship to something, of engaging in any sort of relationship whatsoever with this something.

A notion, then, makes the question prevail: "What is it we are dealing with here?" In this sense, the notion of practice that I have proposed must be appreciated in terms of its relevance relative to the contemporary situation, in terms of what it makes matter, in terms of the questions that it activates. In other words, it should be possible for whoever accepts its relevance to take it up without, for all that, being committed to doing so in a philosophical mode. It is from the point of view of one's own practice, in one's own idiom, that this relevance must be evaluated.

It is a matter now of passing from notions to concepts, because I will make mine the manner in which Gilles Deleuze has presented philosophy, as the "creation of concepts." The philosopher does not claim to be the one who has access to concepts, because for Deleuze one doesn't access concepts; rather, one creates them, and this creation is what philosophy is about. But no one creates in general. One creates because of a question, and to be the "cause" of philosophical creation, that is, of the creation of concepts, this question must oblige the philosopher to think as such. In this instance, it will oblige me to think against what has allowed philosophers to lay claim to the role of the arbiter of common reasons. And it will do this in the name not of higher reasons but the vulnerability that designates me, like others, as a "survivor."

The manner in which I have just characterized concepts—as indissociably and simultaneously what philosophers "create" but also as what "causes" them, as what obligates them to this divergent regime of thought that one calls "philosophical"—is both a borrowing and a capture of a proposition that is at the center of *What Is Philosophy?*, by Gilles Deleuze and Félix Guattari. It is exposed to the risk of being reduced to a "Deleuzean influence." But the question is rather: Why take up this particular concept, the concept of the concept, although Deleuze and Guattari have created many others? The question of influence can then give way to that of problematic proximity, because the Deleuzean concept of the concept comes from a book in which the question, "What is philosophy?" is posed from the point of view of its probable assassination. In this book, philosophy thus becomes what I am calling a "cause," the possible survival of which obliges Deleuze and Guattari to think-create.

Correlatively, it is, in any case, a thinking of divergence that is deployed in their book, because the questions that are also posed there—"What is science?" and "What is art?"—are not "neutral," as if it behooved the philosopher to "define" the answer. For Deleuze and Guattari, these questions are part of what threatens philosophy today. Philosophy is threatened twice over. First, by the generalization of a model of science, wherein questions like, "What is human?" and, "What is freedom?" have to be discussed from the point of view of a possible agreement with and reference to the corresponding state of things. Second, by the flight into the ineffable or sublime associated with a spiritualization of art. It is a matter of separating science from its "logicist" generalization (which is dominated by so many "and therefores" that have the ambition of producing agreement), by associating it with the creation, not of concepts but of functions, and of separating art from a form of piety (which is haunted by a truth that would transcend every "and therefore"), and by associating it with a creation of blocks of sensations, compounds of affect and of percept. Deleuze and Guattari do not define *divergence*; rather, they make what makes for divergence matter.

The concept of the concept in *What Is Philosophy?* obliges one to break with the linguistic habits that render philosophy at once both "sovereign," laying claim to "thinking" what other practices would be busy with, and vulnerable, representing practical questions in a mode that detaches them from those for whom they count but which nothing, as a result, protects from the slippery slope of simple generalization. One often speaks, for example, of the concept of energy, or of space, of time, of matter, most often in referring them directly to a scientific content. Taken in this sense, the concept has, first of all, a reflexive dimension. The philosopher or the scientist himself "reflects on" the manner in which something that was initially a confused notion has progressively been clarified, refined, distinguished, problematized. But in so doing, under the banner of "progress," the philosopher or the scientist generalizes what was an adventure and transforms it into a "moral" development that ought to be able to enlighten every human as such. It is the same, when it is a question of the references to freedom, responsibility, guilt, or truth, that different practices, such as law, confession, or certain forms of psychotherapy, make matter. The intervention of philosophers is destined to gloss over the diverging manners in which these practices confer a signification on these terms, because it seeks to extract a purified version of them, the "concept" that these practices would presuppose and that would belong to the concept of the human as such. Even if this is only to ratify the claims to universality that *cultural studies* denounces.

It is not, therefore, a matter of critiquing the current use of the word *concept*, of prohibiting all talk of the concept of energy, for example, but only of noting that this use does not have any specific relationship with what makes the philosopher, in Deleuze's sense, think. "Reflecting on" corresponds not to a practice but to a general competence, a competence which, as such, is perfectly honorable.

Deleuze and Guattari have written that a concept is not "discursive" and that philosophy is not a discursive formation.[1] This exposed them to a quasi-immediate retort: But what are you doing, if not discoursing, writing books, which are texts like any other, authored, like any other, and belong to the order of discourse, like any other? If they had wanted to produce a public definition of the term *concept*, one intelligible to all, the retort would have been irrefutable, but it is not a matter of a definition, referring to a definable state of things. It is a matter of an attempt to pass on an experience—that is, to *make one experience* what a concept, insofar as it exists, "does" to those who create it or encounter it—for the experience of encountering a concept is what may turn a student into an apprentice philosopher. An attempt to make felt that the concept creates the philosopher even as the philosopher creates concepts, works on and with them, lives through their mutations and rearticulations.

Whom to pass it on to? And how? The question would not be posed if it was a matter of "discursive" statements, answering to the ideal of a "public" intelligibility, which is such that the locutor must be able to answer the question, "What do you mean? Explain yourself!" But it matters, and it matters to me all the more, given that the same question is posed to every practice when it is envisaged outside the common reasons that guarantee its transmissibility to every well-intentioned or well-equipped human.

Thus, to understand a mathematical question is not to have a knowledge of a textual statement; rather, it is to enter into a becoming-mathematician, to experience the efficacy of what nonetheless presents itself as a simple statement. The explanations of a teacher with regard to what a mathematical statement "means" can certainly favor the statement, "I understand!" But they cannot explain it. Likewise, the transmission of the laws and (discursive) theories of a science like physics calls for something entirely different from the elucidation of what they "mean." They aim at making doing [*faire faire*], and Thomas Kuhn has shown clearly the importance of the exercises that figure in physics or chemistry textbooks, without which these laws and theories would have no hold on what they refer to, as well as on those who will make them into tools and who will themselves experience a becoming "entooled."

The idea that a "practical hold" is normal, on the condition that the mind of the pupil—this little idiot who has to be civilized—be emptied of the (idiomatic) "epistemological" obstacles that unfortunately clutter it, is as harmful as statements of the maieutic type, in which it is a matter of giving birth to what the pupil already knows. It's a case of so many denials of a practice, transforming the vital question of its pedagogy into an acquisition that oscillates between formal democracy (everyone can) and a form of elective grace, the math genius or the scientific mind. In both cases, there is a spiriting away of "pedagogy" as a question, thanks to the scarcely perceptible character of its necessity for those who seem to understand straightaway "what it all means." When contrasted with the persistent idiocy of the other pupils, these perhaps not-so-happy few will be defined by their frictionless performance. They will be said to have "the gift," a gift that asks only to be allowed to blossom, like a flower, at the slightest encouragement.

It's not a matter here of discussing the reason why some have more of an ability at math, or philosophy, or. . . . It might raise interesting questions, but these questions bear not on practice as such but on the wrinkles and folds of life that engagement in a practice can give a signification to, but only in a retroactive sense—so that's what it was! In this sense, the evangelical statement, "You would not seek me if you had not found me," is "true," but only for the Christian, whose commitment is conversion, the orientation toward a truth which is that of the Christian soul. As for the blows with a stick that answer most of the student questions in a Zen initiation, they perhaps have a more generic relevance: they translate in the mode proper to this practice the type of "violence" that always signals the pedagogy of a practice. A violence that is radically distinct from that celebrated by the ancient Greeks when they took pride in the substitution of their civilized language for the obscure idioms spoken by "idiot" strangers.

It's this same Greek violence that Deleuze may have been denouncing in his refusal to participate in conferences, public debate, and other opportunities to gather with good will around shared questions. According to Deleuze, the philosopher should act the idiot, or run away, when he hears the phrase, "Let's discuss this,"[2] which often means, "It's my right to ask you all the questions that cross my mind." For instance, I could be confronted with statements like, "It's my right to make you recognize that things exist independently of our practices: This table exists; I can bang on it with my fist. What do you say to that? And don't dodge the question, please." Of course the table exists; that's not the question (here she goes, dodging the question); it's just that for its mode of existence to become a question, one must do more than bang on it with a fist: one might need to take actively into account that it exists in more or less the same way for a cat, which trusts its capacity to resist when jumping on it, but not for a woodworm. Such an answer isn't of the slightest interest to the "discussant," whose sole intention is that of bringing the philosopher back down to earth, of reminding her of common truths. The physicist will, it seems, be more at ease faced with the question of the table's existence: after all it is the physicist who, often, bangs on it with his fist to announce that this table "which we believe to be solid" is in fact constituted of a myriad of particles—the existence of which is, in any case, problematic from a quantum point of view—and a lot of void, which physicists now understand to be. . . . But the table doesn't interest the physicist either: he is "communicating knowledge" and probably is not interested very much in the thorny questions raised by the solidity of materials. He is more at ease here, to be sure, but that is part of the ecological problem posed by the sciences, because in his role as a communicator, he represents "science enlightening opinion," opinion that keeps on asking to be enlightened and doesn't tire of learning that it used to believe, but now has to listen to, what others know.

The most unhealthy situation—which Deleuze and Guattari designated in *What Is Philosophy?* as destined to kill philosophy off—is the one in which philosophers confuse scientific controversies with intersubjective discussions and endeavor to subject philosophy to the norms of this kind of discussion. The statement "I see a table here" then becomes an indifferent starting point for philosophers themselves, on the basis of which one can discuss the question of what it is to exist and, from that point of view, to compare the table, electrons, unicorns, and hypotheses. Have exchanges. Try to come to an agreement. This is what Deleuze designated as the assassination of philosophy.

When I introduced "cause," I introduced an idiomatic term that would not withstand such exchanges. Because I cannot give a definition of it, only explore what happens through risking taking literally what common speech has us say when we say, "He died for his cause," or when Lacanians invented the "cause freudienne," or when a lawyer defends a cause. In all these cases, the word *cause* loses its relationship with a problematic of knowledge. The cause doesn't cause an effect; it is what makes one die heroically; it makes psychoanalysis exist against all odds; it engages the lawyer. It's the same with the word "practice." I cannot define who is and who isn't a practitioner. I have taken the twin signification of this term literally, as what links it to a "doing" (one could pose this question "theoretically," envisage this or that measure, but "in practice . . .") and what links it to a duty (Kant's practical reason), which I translate in terms of obligations. Whoever is engaged in an activity such that "all manners of doing, or of reaching a goal, are not, utterly not, equivalent" is, in this sense, a practitioner. This of course signifies that an economic order in which it is normal to "sell one's labor power" is an order that is destined to destroy practices.

The name of a concept often arises in this way, from a quasi-idiomatic use, diverting a word from vernacular language by making matter a use of it that has not received the power to arm an argument and is, moreover, often difficult to translate into other languages. All philosophical thinking is immersed in the resources of its language. So the trajectory of creation, as well as that of "pedagogy," passes by way of a double engagement: defending idiomatic use against any reductive interpretation, notably that which psychologizes (he believed in his cause), and accepting that the word, taken literally, not as a metaphor for something else, triggers an adventure. It is effectively a matter of following consequences that the initial choice could not have foreseen—that is, of allowing this choice to gain a consistency to the extent that it obliges one to create other concepts that will be components of the initial concept and form its zone of proximity. This—and it does happen—is when the regime of thought that follows a witch's flight,[3] Deleuze writes, can arise, a zigzagging line which is that of the concept as "cause," a line broken up by events when "there is creation." Often it is a matter of almost nothing at all, one more, sometimes almost imperceptible, component or zone of proximity, but a new "perception," in the strong sense of the term, is produced, a perception wherein what is perceived did not preexist, in which the event includes, indissociably, "what" is perceived and "what" perceives.

Philosophy as a "creation of concepts" corresponds to a quite minoritarian version of philosophical practice. The mode of divergence that this creation supposes, which makes philosophers exist and which they make proliferate, cannot be translated into requirements or territorializing boundaries. For many who experience it, it refers to an inadmissible "private" adventure that they have to translate according to the requirements of discursive rationality. And, of course, many of those who call themselves philosophers are perfectly unaware of it. That is why in today's academic ecology, the concept, such as I am presenting it, does not have the power to stir up a community that might defend it as a "cause." On the contrary, one might say that philosophers who have experienced the witch's flight hide it carefully. Deleuze even emphasizes that it is a dangerous exercise, dangerous not only for the one who experiences it but also because of the strong disapproval that it can provoke in public opinion.[4] It is true that in France (not in Belgium) ridicule kills.

It is not without interest to recall here that witches do not attend "conferences" in the mode that "scientific colleagues" invented but gather in "covens," which were demonized, described as the occasion for frightening nocturnal sabbaths. Philosophers are poor things in conferences, mimicking what is appropriate for scientists, subjected to the same academic form. Perhaps the mode of being together of philosophers has something to do with what gathered witches together? What Deleuze sometimes calls "friendship" doesn't have much to do with a friendship between persons; it's much more the knowledge that with this person, it's not necessary to explain oneself too much and that an "exchange" might be possible, like an entangled double flight. In any case, when the exchange has the misfortune of taking place in public, it can horrify. It is a bit like a phase transition, a possession event. People who appeared to be reasonable, well-mannered, seem to fall into a delusional bout of jousting: it's as if no one else existed. Then they come back down to earth and make apologies.

We may be touching here on the risk associated with this apparently abstract story of cause. Because the objection is ready to hand: Aren't you describing what we call a "sect"? And that is indeed what it's about: not the defense of "sectarian thinking," to be sure, but the fear that "secret societies" arouse. As a philosopher, I am under much greater threat than others from the accusation of practicing a language that is "sectarian"—a language that is in fact opaque, although it seems to use the words of everyone. So I am doubtless more sensitive than others to the manner in which the public order has exploited this fear and has been maintaining it so as to translate into a quasi-moral imperative the act of having to explain oneself, of having to speak a public language, one accessible by rights to everyone.

The stake of the second part of this book is to try to address this fear. It is impossible for me to dismiss the possibility of a "war of causes" as an illusion that would be dissipated in the light of an ecology of practices, to pretend that practices can only be "good." But the fact that peace cannot be guaranteed doesn't signify that this other modality of war, called "pacification," the destruction of what does not submit to so-called common reason, is legitimate. To think in terms of an ecology of practices is to think in a hand-to-hand combat with epic or missionary rhetoric, which gives its blessing to the destruction of what creates trouble or fear or obstacles, justifying wars that are waged in the name of the necessary understanding among well-intentioned people.

ON THE SAME PLANE?

But That Would Open the Door to . . .

We need to be able to hear the protest that if it's not just worthy candidates but also unreasonable activities, such as fortune-telling, magic, astrology, that we put on the same plane—that of practices—the door will be open to charlatans, to imposters, to creationists . . . to no matter what. That philosophy works within divergence doesn't mean that it can do without the question that such a protest poses, because, since Plato, it is this protest that has linked the practice of philosophy to defending public order. If the test that the protest poses is not accepted, then every ecology of practices that doesn't respect common judgment (the opposition between the Virgin Mary and the neutrino) will not just be disregarded but, worse, will confirm that calling into question the role that is conferred on scientific knowledges and on the policing of frontiers leads directly to the most absurd, the most worrying of consequences. Accepting the test of this panic, this Dostoyevskian protest—everything would be permitted—is one of the consequences that gives to "thinking through causes" its demanding signification.

However, accepting the test of hearing this protest does not in the least signify accepting what impels it, that is to say, the image of a quasi-demonic mass of delusions, fanaticisms, monstrosities, credulity that push at a door that can only be kept shut by the heroic effort of the white man and his requirement of rationality. The slightest weakening, the slightest sensitivity toward the sophist sirens who threaten to complicate the question, and the confused, ignominious mass will smash the door down and invade the public stage. To accept the test is to know that one is exposed to the imperious calls to order that invoke this image and not to be scandalized by them, as if they stemmed from some sort of misunderstanding.

If I mention imperious calls to order here, I do so in order to emphasize that those who protest, "But that would open the door to . . . ," do not themselves personally feel threatened by the monsters supposedly behind the door. It is a matter of protecting others—more precisely, those who have been assigned the role of being edified by the public representation of science. It is primarily a matter of protecting them from themselves, because it is inside them that the monsters behind the door fester; they are the ones who would make the violence of unbridled passions reign, if they emerged from their passivity.

Many histories can be recounted regarding this image, which nowadays plays the role of an order-word. Many dates can be lined up together. One can, of course, go right back to Plato, opposing to the sophist Callicles, who claims to seduce the crowds so as to dominate them, the authority of a rationality that transcends the passions of the people. One can invoke recent experience: Nazi delusions, sects, religious fanaticism. . . . Between the two there is doubtless the enormous fear of the bourgeoisie faced with revolutionary crowds. But prior to that, there is also an episode that the historian Robert Darnton has associated with the "end of the Lumières," the end of the alliance of those in the eighteenth century who were called the "philosophers" and a "public" that, like them, was avid for knowledge and rebelled against traditional authorities. In the years before the French revolution, scientists discovered, with fright, the "credulity-of-a-public-that-can-be-duped." Plato was right: the Callicles who would seduce a seducible people was named Anton Mesmer.

Around Mesmer's *baquet*, filled with a curative magnetic fluid, women fainted; some surprising cures were produced; and the crowd became impassioned. In 1784, academic scientists, members of a committee of inquiry nominated by King Louis XVI, devoted themselves to combating this troubling of public order. In order to do this, they would adopt a role that can be likened to that of the *testator* in the service of a prince, testing the gold that alchemists would offer to the prince who paid them. These scientists put Mesmer's fluid to the test and concluded that it could not claim to have a scientific mode of existence, because if one separated it from the "imagination" of those it was supposed to affect, it had no effect. Mesmer was therefore a charlatan.

The role of *testatores* assumed by the scientists made the capacity to pass the test, to resist competent objections, shift from the register of the event to the general register of judgment. When *testatores* are at work, there is no "success."[1] The scientists on the "Mesmer committee" did not articulate Mesmer's cure in a mode that would allow its effects to be interpreted in a reliable manner. To dismiss these effects as merely due to the subjective beliefs of those whom Mesmer had duped, it was enough for them to show that Mesmer's fluid was incapable of claiming experimental reality. Science here is defined as being *against* opinion; the requirement of proof arms a general tribunal that, above all else, is intended to establish a pedagogic relationship with a public henceforth defined as credulous.

It's not a matter of denying the problem of credulity, of dismissing it as an illusion, and of proposing the counterimage of an angelic and lucid "citizen" public. Rather, it's a matter of thinking on the basis of the ecological disaster constituted by a situation in which proof, which only "science" can provide, has become an order-word intended to protect the public. Proof becomes a required duty. Even politicians are now treated as demagogues if their arguments are not based on proven "fact": the power of proof has become something that every protagonist on the public stage must be able to lay claim to.

To multiply stories and dates is to create the active memory of a succession of events, while the "order-word" presents self-evident continuity. It is perhaps also to make felt an affect—contempt, fear, fright, disappointment—that is activated every time a thinker announces, as an irrefutable objection, "But that would open the door to. . . ." However, stories, questions, and arguments can't do much against an order-word. The order-word feeds off what it produces—in this instance a throng of pseudoproofs, "charlatans" who assert that their claims are "scientifically demonstrated."

"How can one not hunt down such charlatans?" one might ask. Yet it should be emphasized here that such a hunt is selective. Without mentioning the relative respect that economists benefit from, it is striking to note the extent to which experimental psychologists (respectable) and "parapsychologists" (hounded) now resemble each other in being only interested in situations regulated by the possibility of a statistical "monstration"[2] that stands in for proof. The only difference is that the "proofs" provided by the former accumulate in the benign indifference of the members of other scientific communities, while the latter see theirs pulled to pieces by suddenly lucid critics. And the difference between the experimental proof and the statistical monstration is completely forgotten by the medicine that grounds its authority on clinical tests in order to present itself as "based on clinical data."

Excluded from all this is the public, which is really at the center of the contemporary ecology of scientific practices, because it is always the public's worrying credulity that is the argument used to reject questions one suspects might interest them. At the limit, this last argument could be accepted by those who object, "But that would open the door to . . . !" But they will still continue to object, on the basis that, "Unfortunately there isn't any other solution!" The question of a possible ecological transformation thus demands that one not come to a halt with a mode of description that "deconstructs" fright while leaving it intact, thereby provoking the hateful acrimony of those who risk becoming imprisoned in a, "We know very well, but all the same." The possibility of putting what are normally opposed—the Virgin Mary and the neutrino—on "the same plane" is apt to engage philosophy in a conceptual creation, but this latter cannot abstract away the political question that provoked it. If not, the conceptual adventure would be suddenly transformed into an "epistemic sovereignty," which would allow the philosopher to pass judgment on those whom she is defining in terms of a fright that she herself does not share.

It is thus important to underline that "I" am not someone who doesn't share the fright that I diagnose in "others." I accept that, "like others," I am vulnerable to it. This is the problem that engages me as a philosopher and that obliges me to define as a failure anything that could lead to philosophical concepts being transformed into pedagogical instruments designed to bring safety and destined to ratify consensual judgments as to what is and what isn't acceptable. It is what makes me risk a different "version"[3] of what can, in effect, provoke fright.

To think in terms of an ecology of practices is to try to displace the fright, to stop being frightened by people's credulity, but to be so instead by the contemporary ecological assemblage, populated with predators and their prey. It's a matter of ceasing to think first about those who belong to this public, whom we don't know and about whom no one today can say what they are, or might become, capable of. We know only that the manner in which they are excluded by the ecology of predators and prey, protected like irresponsible children, and kept at a distance from what nevertheless concerns them actively contributes to what it is that makes them frightful. We have no guarantees with regard to a different future, a future belonging to no one. But by contrast, it does behoove us to stop defining ourselves as guardians of this future, as being in charge of those who are not able to think for themselves. Because if this responsibility poisons us, if it contributes to transforming the practices that make us think, feel, imagine, and act into norms that consign us to fatal submission, it is "for us ourselves" and not "because we trust them" that we have to discard this responsibility, to change the problem, to dream other dreams. To experiment with other versions telling other stories.

It might be said that this is pure *fabulation*. But fabulation is not a hollow dream if it succeeds in creating the appetite that alone is able to defuse the panic reaction and its "I know very well, but all the same. . . ." What would be a hollow dream or badly utopian, by contrast, would be the idea that if it was included and not excluded, the public would suddenly become reliable, lucid, marvelously ready to belie the fright it provokes, as if by the grace of the kiss that transforms the frog into Prince Charming. Fabulating an ecology of practices is an active fabulation, because it concerns what those to whom it is addressed might become: it asks them here to give up the demand for reassuring guarantees about the future, to break with a (bad) history that turns certain practices into the brains of this future. That's why such fabulation is political, because politics, in the sense that matters, begins not with a people that is finally reliable but by abandoning the defenses identifying that people as unreliable, defining it as irresponsible. For everything else, as Whitehead wrote, "*It is the business of the future of to be dangerous.*"[4]

The Truth of the Relative

To fabulate, to tell different stories, is not to break with "reality" but to seek to render perceptible aspects of this reality usually considered accessory, to give them the power to make us think and feel. And it is above all to escape from the "fable of origins" that we have had to recount since Galileo. According to this fable, fabulation is nothing but fiction, free, for sure, but primarily arbitrary. It translates the power of the "whys" that haunt human subjectivity. The truths that it seems to produce are relative to our ideas, to our convictions, to our habits, not to what we deal with.

Since Galileo, many human sciences have transformed this fable into a secure starting point. Thus, one might say that what a fabulation aims at arises from the "performative character of language," its capacity to make exist what it seems merely to describe. A very interesting proposition, certainly, if it wasn't intended as a secure starting point for "showing the facts." Because the consequence of such a starting point is that the performative characteristic is usually associated with what allows an adherence or illusion to be explained, not with what makes one think or imagine. In other words, what gets evacuated here, to the profit of a general mode of explanation, is nothing other than what is at stake with the fabulation that I am proposing, the *creation* of a different manner of perceiving, of being affected, and of being frightened. It's the same with the different "effects" identified by psychosociology, which demonstrate how much what is thought and perceived can depend on the manner in which a situation is staged. In a significant manner, these effects bear the name of experimenters: the Milgram effect, Sherif's autokinetic effect, the Rosenthal effect, the Valins effect, the Hawthorne effect, Lewin's freeze-unfreeze model, and so forth. . . . It is in effect a matter of behaviors prompted by the apparatuses these experimenters invented, and the feature common to all of them is that they almost inevitably place "subjects" in a situation of inferiority—victims—because in order to be able to "show," the scientist must be in a position to differentiate between the "true" situation and the manner in which its subjects behave as a function of its staging. The behavior is thus judged to be a "function" of the staging. The idea that what is produced "exceeds" the experimenter, that is to say, that his subjects become capable of something that he himself is not, is excluded.[5]

John Dewey's complaint, according to which the "social" sciences do not have to explain "the social" because association is the mode of existence of everything that exists, can be generalized here. In addition, the "social" sciences do not have to explain the associations that, obtained through deliberate manipulation, they have given themselves the means to judge. What do we know about what the manipulation has produced? Have "subjects" been constituted as the "guarantors" of their own reaction, authorizing the affirmation that the latter is indeed the effect of the manipulation?[6] With Dewey, and beyond Dewey, I will affirm that the creation of associations is the primary fact, producing its own terms. And all who set out to dismember such associations in the name of their science, to deny the event of creation so as to refer what makes people think and feel to the human, always to the human—whatever their good intentions might be—must be characterized on the basis of the formidable passion that animates them: destroying, dissolving, detaching. Deconstructing. *Reducing the truth of the relative to the relativity of truth.*

That the relative has a truth is an important proposition of Gilles Deleuze and Félix Guattari's,[7] and a proposition that is crucial for my enterprise of fabulation. The truth of the relative has nothing to do with the relativity of truth, which—explicitly or not—stages a position of judgment, able to define what presents itself as truth as "only relative" to (convictions, social membership, class interests, culture, and so forth). For its part, the truth of the relative affirms the indissociable character of truth and the creation of a relation, without any reference to a "detached objectivity," whether it is reputed to be possible or not. The relation doesn't explain the truth; it creates the terms between which it will be a question of truth. Fabulation attempts to create such a relation, productive of new manners of being affected, of feeling, and of thinking.

Homage here to a fabulating sister, Donna Haraway, too often transformed, in spite of herself, into a heroine of postmodern relativism in its struggle against modern "grand narratives." When Haraway presents herself as a child of the Scientific Revolution, of the Enlightenment, and of Technoscience (the science that patented OncoMouse, the mouse genetically modified to "make" breast cancer tumors), she deliberately sets out to construct other modes of narrative than the master narratives of progress. But it's not a matter of "deconstructing," in the sense of reducing to fiction—a Galilean gesture. Stories are not "simply" fictions. Some stories, those that prevail today, continue to make us the children of Galileo or Boyle, whereas we now live in a world in which mice are patented. The stories that Haraway constructs are tools for new manners of situating ourselves, imagining and thinking.[8]

As it happens, Haraway also breaks with the division that eminent scientists rely on when they judge that they are free to recount histories that will keep the public at a respectable distance: there is the properly scientific domain, in which the strictest demarcation must be maintained between what can be proved and ongoing, still-uncertain research, and there is everything else, which is nothing but fiction, where these scientists are free to recount the history as they feel appropriate, to linger over promises, and to pass over what they demand. By learning how to narrate this division itself, Haraway certainly undermines the distinction between "science" and "ideology," but she doesn't dismiss everything as ideology. On the contrary, the stories that she has learned to assemble have the "truth of the relative." They do not reflect an ideology, the generality of which would subject every case to the same point of view. They participate in the creation of a differently situated "witness," endowed with a new point of view, which is susceptible to being affected by the sciences in a new mode.

OncoMouse has the truth of the relative. It functions in Haraway's text as the incarnation of and the witness to natural-cultural traffic, "transgressing the frontiers" that were officially decreed in the epoch when competent colleagues gathered around Boyle's air pump. It renders perceptible the inextricable links that are fabricated between living and nonliving matters: human suffering and mouse suffering, genes, research, patents, and industrial strategies. Whereas the scene of the past, officially depopulated, purified of everything other than competent colleagues, has been susceptible of invasion by interested protagonists; OncoMouse cannot filter legitimate and illegitimate interests, nor can it denounce what is responsible for its existential suffering. It suffers "for women threatened with cancer" (but which women?), and it is not equipped for denunciation. It can only ask that one learn to perceive the past as much as the present in a different way. To perceive past and present in a mode that provokes care for the difference established between those who participate in the production of knowledge and those who have to be content simply with looking or those who, like OncoMouse, are destined to suffer. OncoMouse then resonates with the birds who died in Boyle's pump, so as to demonstrate that what was evacuated was necessary for life.

"Deconstructivist" readings of Haraway's work have most often ignored her care, so as to celebrate heartily a world of transgressions, in which all identity is factitious. Such readings wouldn't have any problem deconstructing the care that I am continuing to call for—the care not to insult practitioners—by interpreting it as what prevents me from accepting that what I have called "cause" is nothing but a disguised manner of defining what unites practitioners as having primacy: "their" cause, which they defend. Hence, the relativity of the truth: what practitioners do is relative to what is important "for them." The peculiarity of the "truth of the relative," however, is to be vulnerable to such deconstruction, not to be armed against it but to appeal to the creation of perceptions that resist it.

It's a creation of perception of this kind that I am appealing to so as to fabulate scientists who, instead of denouncing the irrationality of sociological analyses that "demystify," would protest that "the manner in which you describe us, as addressing a reality that is incapable of differentiating between us, is destroying us! We are no longer anything without an association with this nature that you are characterizing as mute, as radically incapable of playing the role that we appeal to it to play, that is to say, as incapable of making the possibility of our agreement anything other than a human convention!" A scientist able to protest in this way would have become able to participate in an ecology of practices without playing the role of a predator in it.

CHAPTER SIX

Today, only a few mathematicians—those, notably, who are not worried when affirming that they are Platonists, that is to say, obliged by the mathematical beings that make them think—are able to smile gently, whereas the physicists get angry. They know, and they know how to honor, the association that coproduces mathematicians and their beings. And the idea that the new truths that they produce are created in the relation, and so are inseparable from new manners of perceiving and being affected, wouldn't be a scandal for them.

To fabulate an ecology of practices is to fabulate practitioners able to smile like mathematicians, having no need to defend themselves by having recourse to the general values (rationality, proof, facts) that arm an ecology of predators and their prey. But the possibility of such practitioners would remain in the order of a pleasant fiction—and not a fabulation that opens new appetites, that displaces questions—if it concerned only scientists. It would leave intact the primacy of the stakes for knowledge that arms predators. That is why fabulating an ecology of practices that is not reducible to an ecology of predators (judges) and prey (those judged) requires that the ensemble of practices become capable of refusing to be judged in general terms, reduced to a "purely human" matter.

It's easy enough to imagine that, like the mathematician, the pilgrim en route to the Virgin Mary might smile at the reduction of her experience to an "effect" that can be analyzed in purely human terms. Pilgrims know that they are living in an "unhealthy" world, which tolerates them, which defines them as living relics or as members of a species on its way to extinction. They are not in the position to protest publicly against descriptions that give human subjectivity primacy and that define pilgrimages as simple, private matters, matters that are no more or less respectable than undergoing psychotherapy. The pilgrim's smile testifies to the irrelevance of these descriptions, but it cannot, as such, make specialists, in reducing everything to the human, hesitate, because it has been situated and judged since "the "origins" of modern science: science against belief.

This is why, by way of an example, I turn here to an ancient practice, one which, like philosophy, has not only survived eradication but has also been "modernized": the practicing of the law, which, as Bruno Latour has demonstrated superbly, is irreducible to a general convention.[9] I won't take up here the set of requirements, described by Latour, showing what a "well-made" judgment has to satisfy. What I would like to underline is that this fabrication of law is vulnerable to being reduced to a "purely human" matter, in the manner that the "truth of the relative" is vulnerable to being reduced to the relativity of the truth.

So then the question is this: And if what the reduction of law to a purely human matter ignores was precisely that which obliges the law's practitioners to hesitate in the mode that signals a practice?

When it is a question of the fabricating of judgments, practitioners are not vulnerable any more than scientists and their controversies are. There is effectively hesitation, but the intruder who would introduce an, "All you have to do is . . .," would find himself thrown out with as much determination as someone making a suggestion of the same kind in a laboratory experiment. And yet I would propose that what makes law a vulnerable practice is the question played in this hesitation by the reference to what is "just." Here, the relativistic "intruder" has a certain capacity to be harmful. He will sneer that what is just is nothing more than what the verdict that has satisfied its legal requirements pronounces. And practitioners would be vulnerable to this sneering. They know very well that their practice is not subjected to "justness," is not explained by what is "just." Most would nonetheless be happy that the possibility of speaking of a "good judgment" be acknowledged, as when opposed parties can admit a judgment's pertinence, even if they end up being disappointed by it. But the words available to them in this regard designate only the "human quality" of the judge, and as usual this psychologization signals mainly the avoidance of the question, the fact that it is not cultivated in the training of lawyers.

In *What Is Philosophy?*, Gilles Deleuze underlined the importance of a "pedagogy of concepts" that might make one feel their difference from discursive propositions, which allow for an assured linking of "and therefore" and "so." When it is a question of the concept, an "and therefore" is always an adventure that can make the problem collapse, leading one, as Leibniz wrote, back out to sea, when one thought one had reached port. And the reason to which "so" refers can, by this very reference, by this intervention in an argument, cease to be authoritative, opening up new questions. A pedagogy of the trajectory that produces an "object of law" from an initial situation of conflict, through the arguments of the parties involved, to judgment might be equally necessary for practitioners to be able to affirm the truth proper to their practice, the art of a fabrication that has nothing to do with the satisfaction of formal requirements, because it is the judgment that defines both the object and the manner in which these requirements impose the "and thus" of the decision. And who knows? Perhaps then these practitioners might respond to the relativists: Sure, we don't have any privileged access to "justice," but our labor is to unfold a situation in such a way that the maximum hesitation is imposed as to the manner in which it lends itself to judgment. It's thanks to this labor that judges can resist the cynicism and routines of the power that they exercise and that is exercised over them. The truth of the relative.

Productions of Existence

With the example of the law we have moved away from the ensemble of practices, whose stakes might be said to be "learning something new" with regard to the world. A decision about justice is of course an event; it creates a "before" and an "after," but its value doesn't derive from its newness. The law, Latour has proposed, is a form of antifiction, if fiction is what permits us to transport ourselves into another time, another place, to experience what it is to become someone else: a princess, a drug addict, a banker, a Tyrannosaurus rex, or an ant. Law attaches, authenticates, assigns, links, traces, gives "legal" consequences to an action, a signature, or an agreement. We are not the "subjects of the law," but the law makes us the subjects of its utterances from the moment our actions generate consequences belonging to its register. In other words, the law makes precise, assignable consequences exist for something that would otherwise remain a matter for indefinitely contestable, undecidable interpretations.

If there is a trait that is common to both scientific and legal practices, this trait doesn't bear on something that would be a property of the human in general; it can in no way be derived from general faculties of knowing or judging; it doesn't translate questions about what we can know and what we are capable of evaluating. It refers instead to the question, which is never general, of a *production of existence*, a "making exist," the mode of which is immanent to each practice.

I have already initiated a transition from the question of knowledge to that of productions of existence, and I have tried to make perceptible how words such as *interest* or *importance* change meaning when they are taken up in this transition, separate from the register in which they signified a subjective value likely to put the objectivity of knowledge in danger. But the moment has come to attempt the "great leap" to which the statement "production of existence" incites. Because if there is a domain in which "interest"—in the sense of *inter-esse*, or a bond that articulates—importance, and production of existence are irreducibly coupled, it is indeed that of the living being.

The question will not, of course, be to produce a "general theory" of the living being, which would come to involve the question of practices, or to compare a practice to a living being. That would be a vision of the world, a flight that soars through the space of metaphorical generalities, whereas concepts demand an exercise in "literality." It is thus a question of seeing how living beings, through those who study them, literally testify to the type of force that I am trying to confer on practices, that of making the dramatic alternative between truth that transcends opinion and the relativity of truth stammer.

Let's take the contrast between the first bacteria, anaerobic bacteria whose metabolism produced oxygen as a "waste product," and the "newer" aerobic bacteria, which changed oxygen from being a poison (if found in too great a quantity in the atmosphere) into a vital resource. No self-respecting biochemist would ever think to claim that this innovation manifests the contingency or relativity of any definition of oxygen. She will instead celebrate the amazing feat of a poison having been transformed into a resource, a new biochemical signification having been conferred on the properties associated with oxygen by chemistry. She will study the detailed set of metabolic assemblages that associate subtle and varied consequences with the new relation created with oxygen. Similarly, when the movements of protowinged creatures made the new mode of interaction with the air that is called "lift" exist, no bioengineer would say that the manner in which the air matters for species that explore the possibilities of what is now an "aerial milieu" translates the relativity of our definition of the air as a factor of friction. In both cases, innovation is inseparable from the coming into existence of new beings adventuring in a world of new possibilities and risks: truth of the relative.

The history of evolution abounds in such mutations, and in each case, what we call "interest" and associate with humans can be used to correlate with "what is situated between," what brings into, what creates, a relationship—on the condition that its etymological roots are reactivated. It is not enough to say that the relation between living beings and oxygen, or between the movement of a living being and air, has changed. One really must speak of the creation of a relationship. There is effectively a difference between oxygen-poison and oxygen-resource, just as there is one between air as a friction factor and air that plays a role in lift. As a poison for the living being, oxygen can be explained in general chemical terms, by means of its oxidizing properties. Conversely, although the general properties of oxygen are certainly required by aerobic metabolism, the connection that metabolism brings about between these properties and the vital stake constituted by energetic metabolism have nothing general about them. Similarly, as a friction factor, air is a generality that concerns all moving bodies on the planet Earth, whereas the lifting role that air plays implies a bringing into relationship the variations and perfecting of which will open up a field of extraordinary experimentation, putting into play the assemblage of feathers, muscles, bones, and brain.

In both these cases, one can say that what biologists usually call "adaptation" doesn't have very much to do with the injunction to "adapt!" or with "this is how the world is; deal with it!" but much more with the traits of what William James called "verification." The novelty of a relationship brought into being is "important" to the extent that it confers on the multiplicity of "details" a power of counting, of making a difference, in short, of becoming interesting. The great variety of modes of flight today—from crickets to bats, via swallows—cannot be deduced either from the air or from lift, as the engineers have been able to define it. On the contrary, engineers and biologists are themselves passionately interested in the extraordinary feat that each of these modes constitutes, the manner in which each selects or neglects aspects of a world that thereby becomes more and more concrete, more and more interesting.

It is not enough here to emphasize that in order to survive, every animal makes certain aspects of its milieu matter in a mode that is specific. Karl Popper often insisted on the fact that all living behavior is an "embodied" hypothesis that bets on a milieu that isn't in the slightest a general environment but the environment endowed with what the hypothesis in question differentiates in terms of value. But it was also Popper who stated a principle of demarcation that put deliberate risk at the heart of scientific innovation: "The amoeba dislikes to err while Einstein is intrigued by it: he consciously searches for its error in the hope of learning by its discovery and elimination."[10] The positive risk of refutation is then at the center of intellectual life, which is thereby opposed to the stakes of animal survival. And it is here, of course, that it is worth slowing down, because, like Dewey's experimental logic, demarcation goes much too quickly. It turns the particularity of experimental practices, which link the value of a hypothesis to the possibility of it overcoming the most demanding (potentially lethal) of tests, into a privilege, and it opposes these practices to everything else: from amoeba to everyone who, like amoeba apparently, acts through conviction or habit.[11] An ensemble that is ruled by a generality, the stake of survival.

The question of how to "describe well" animal, even vegetal or micro-bial, modes of existence pertains to biologists, and many cultivate it admirably. But biology is also subject to the power of generalities, through which it is transformed into a reduction machine against which the human will have to prove his privilege. If one agrees to abstract out the question of "how" animals satisfy the "needs" that can be assigned to all living things, if one confers on these needs the power of explaining animal "mores," nothing will be able to stop the assimilationist machine. And one will arrive at Jacques Monod's celebrated statement, according to which a poet's talent is perhaps explained by the necessity of recruiting a female in order to be able to transmit his genes. Or the viewpoint of evolutionary progressivists narrating how the needs of humans have always been best satisfied by the division of labor, the creation of cities, governments, nations, and so forth. In order to refuse that all human production can be captured by this common judgment, some will come to oppose, in the style of Gaston Bachelard, the "interests of the mind" and the "interests of life."

To resist, here, is to resist everywhere, but on a case-by-case basis, never in general. It is to pay attention to each new mode of perception, to the invention and taking on of consistency of each new type of connection, setting living beings off on adventures that, from the point of view of "needs," are "crazy," fabricating a world that is always extravagant and prolific. This is how Gilles Deleuze and Félix Guattari proceeded, in *A Thousand Plateaus*, making concepts proliferate everywhere, against any possibility of subjecting living beings to generalities.[12]

From the point of view of productions of existence that I am trying to construct, it is their concept of territory that I will retain as crucial. Against Konrad Lorenz, who "invented" the territorial animal but situated it by setting it in an adaptationist continuity which acts as a vector for every naturalist reduction ("we are aggressive because we are territorial animals"), Deleuze and Guattari thought the invention of territoriality as irreducible to an adaptation that is explained by the same imperatives as those that would explain the behavior of nonterritorial animals. To be sure, like all other animals, territorial animals—those for whom the contrast mine/another's has a sense—have to feed, to reproduce, to escape from predators. The general imperative organized around the axis of differential survival is no more contradicted by territory than the general chemical properties of oxygen are contradicted by aerobic metabolism. But it remains the case that Lorenz misses what matters: while the general imperative to have to reproduce might seem to have the power to explain, when it is a question of parasites, for example, satisfying this imperative is required only for territorial behavior. It is incapable of explaining their fabulous divergences, the multiplicity of what they make matter. The "invention of territory" changes everything, for it is on this basis that a multiplicity of new adventures will unfold. Truth of the relative.

For Deleuze and Guattari, territory and the adventure of deterritorialization are in a situation of "reciprocal presupposition." It is with the meaning of "at home" that the experiences of "leaving," "exploring," "venturing outside" take on meaning. And this is where the difference between what they call "relativity of truth" and "truth of the relative" is played out. Because to take territory *and* deterritorialization seriously is to affirm a distinction that is not only "observed" by the biologist (I observe that sunflowers turn toward a source of light) but also has a meaning "for" the animal. It is difficult to share the adventures of plants, but what the animal who ventures into the territory of another experiences is neither subjective nor objective, but has the truth of the creation of a connection that is invented together with the territory. Animal experience itself becomes the witness to a world that is populated with new beings that are enrolled by these creations of relations.

Ethologists have, for example, observed that the time and energy devoted by certain birds to singing seems considerably greater than what would be required in the interests of defending a territory or attracting a female mate. From the adaptationist point of view, these birds are crazy! Can we say that their singing testifies to a new kind of enjoyment implying a new "relationship to self"? When birds sing at sunrise, can we dare to say that they are giving a meaning and a reality to this "sunrise" as an event that matters, as innumerable populations of humans celebrating the celestial body also do? Wouldn't the enjoyment of the songbird, the existence of the songbird as joyful, itself be the creation of a hitherto unknown bringing into relationship conferring new roles on the vibrating body and sound, making the event that we call rising-of-the-sun come into existence?

This is the point at which it will be protested that the Sun existed well before birds celebrated its rising! The Sun's reality is perfectly indifferent to the appearance of living beings, for whom its "rising" has meaning, even importance. Its light affects the sensory receptors of animals independently of the territorial/nonterritorial distinction. While the examples of oxygen and air might have worked, because lift and the biochemistry of aerobic metabolism are authenticated by "objective" knowledges that describe the roles played by air and oxygen respectively, here there's no getting away from the challenge of the "truth of the relative." Between the Sun, which as we know "rose every day" well before there were living beings for whom this event matters, and the sunrise of the bird's song, the difference between objective reality and subjective importance lashes out. Even if this difference means forgetting that the objective definition of the "sunrise" is itself inseparable from the history of generations of astronomers all over the Earth for whom this event also mattered.

We are not done with "objective reality," with the importance that we confer on a reality whose indifference guarantees its independence in relation to our fictions. It's precisely because the bird, like the plant, doesn't only "need" the Sun but testifies "for" the Sun, that this "objective reality" makes its return, threatening to dismember the site where the concept of "cause" such as I have proposed it was taking on meaning: sunrise as "cause" for the bird, what makes it feel and sing.

The Powers of Reality

To fabulate is also to hear a cry of protest, which it matters to "comprehend"—in the sense of "taking with"—to bring into the fabulation, rather than to silence. The cry, "But the Sun existed before!" translates one of the important meanings that we attach to the complicated term *reality*. "Reality" imposes itself independently of the bird: it is only the bird that attributes a purely avian signification on it, without that making any difference to the Sun. To speak of the "creation of a relationship" between a virus and an avian species is acceptable, and the possibility of the avian flu virus becoming capable of creating a relationship with humans is one that is feared. But it is supposed that the Sun itself is not sensitive to the entirely relative event that its "rising," observable from some point on Earth, constitutes. To speak of a bringing into relationship would thus be sophistic wordplay or simple metaphor.

However, in order to confer the truth of the relative on the rising of the Sun, there's no need to maintain the Aztecs' conviction that the Sun must be nourished. With regard to their "new perception," marked by a sense of being "in charge," one can speak of a capture that is "unilateral," that isn't negotiated but that really does confer a role on what the Egyptians for their part celebrated as the source of all life. The Sun is enrolled by those living beings that count on its light or warmth, by birds that sing, by humans who celebrate it, and by scientists who oppose its "objective" reality to the subjective perceptions that we have of it. The "objective" reality of the Sun of Newtonian astronomers, which attracts the Earth but which is also attracted by the Earth, is the fruit of an adventure that has succeeded in conferring on the Sun and the planets the power of making a difference to the very manner in which we describe them, in spite of their indifference.

Let's take another example of the "reality" to which we attribute the power of imposing itself independently of our perceptions and projects. We do not know what those people who suddenly saw the massive tsunami breaking on the morning of December 26, 2004, were busy doing, what they were hoping or feeling. Whatever the case may have been, it is pretty much incontestable that the wave had the power to erupt, to interrupt, to dismiss as insignificant whatever was occupying those people moments before. The power associated with this wave is a power to intrude and interrupt that we can describe as unilateral, indifferent to what it interrupts and destroys. In any case this is how "those who know" will define it, asking all humans to have the courage to recognize that our intentions, our projects, what we trust in, are like fragile skiffs that can sink under the blind tumult of a tsunami for no reason of their own. And for most, this demand entails an adhesion that marks the manner in which something that in the past was a difficult and demanding thought—that of the first atheists or the Stoic *amor fati*, the love of a destiny indifferent to this love—can be transformed into a habit of thought.

But a rather remarkable slippage of meaning occurs here, because this demand, and the adhesion it provokes, now appear as "normal," identified with anonymous progress in opposition to the particular "beliefs" of those who seek intentions everywhere, who cannot admit the "absence of reason." However, rather bizarrely, those who know that it is not worth looking for a reason are also those who frequently seek those responsible, who inveigh against the absence of a dense and rapid network ready to establish within minutes the connection between the automatic registration of submarine tremors and an alarm signal received by every inhabitant of every coastal village. Why, they ask, at the start of the twenty-first century, is such a disaster still possible?

It is not a matter of dismissing, as the same, traditions of thought for which there is no "natural," unintentional, interruption of a life and those that both affirm and refuse the hazardous nature of things, who glorify in not asking "why" when it is a question of what is called nature, but who monitor with a persnickety attentiveness all the "whys" that relate to human choices.[13] What has to be underlined here is that this undeniable reality—the tidal wave that interrupts so many lives—mattered to all of those that it affected, and no one doubts its "reality." But it does not have the power to bring everyone into agreement. Not those it affected directly, some fleeing, others remaining, petrified, or trying to protect their children, nor those who, like elephants, seem to have "known" . . . or those who now discuss the consequences or those who, through prayers and rituals, seek to give a meaning to what hit them. But it is precisely this register, of the power of bringing into agreement, that is referred to by what is called experimental, and more commonly, "objective" reality, which I have taken the neutrino as a representative case of since the start of this book. A neutrino has never interrupted anything at all; it is even characterized by a power to intrude that is so slight that detecting it is a major experimental achievement. Yet physicists have conferred on it the power to intervene in their discussions, to arbitrate in their hesitations: it does indeed, "objectively," have a mass. It really exists!

No one will say: the murderous tidal wave "really existed." Such reduplication is meaningless, because it pertains to the intruding and interrupting power of the tsunami to suppress any doubt in this regard. It is the same with the lioness who suddenly attacks the careless hunter or with the rain that ruins picnic plans. By contrast, it can be said of neutrinos, because their existence imposed itself against a background of doubt. To be sure, the real existence of neutrinos didn't silence public opinion, which really couldn't give a damn, but, rather, the opinions of the competent, of everyone who subsequently suspended the solar neutrino anomaly and who rejoiced, like Karl Popper's Einstein, over the death of their respective opinions.

Here, those of you who have understood that I am in the process of undermining the holy alliance between reason and reality will cry "hypocrisy!" This philosopher is in the process of constructing a get-out clause again! She knows very well that it was a faultlessly recorded, identified, measured, seismic event that caused the tsunami! She is in fact using the same rhetoric as those deconstructionists who maintain that the laws of physics are simple human conventions and who, in order to explain why, in this instance, planes don't fall out of the sky, propose that we put the prayers of anxious passengers and the laws of physics on the same undecidable footing! Would even the most devout of passengers trust an airline that put its planes in the hands of prayer groups and not competent technicians? To this I would reply that I would no more trust this airline than I would one that put its planes in the hands of physicists who treat flying planes as illustrations of the laws of nature. This reply is not a joke; it points toward the manner in which the sciences represent themselves publicly as based on "reality" and as "explaining" techniques. But the objection must nevertheless be heard: am I free to play the naturalist, who can place the extraordinary diversity of manners in which milieus count for living beings on the same plane—that of wonder—while of course I am among those for whom the murderous tidal wave created "innocent" victims, that is to say, wasn't itself animated by any punitive or vengeful intention?

The objection seems to me to make the difficulty of thinking the truth of the relative vibrate. It makes what we know of the link between the tsunami and plate tectonics intervene and refuses the possibility, judged scandalous, of putting this link on the same plane as other interpretations of what happened, including those that make a form of intentionality intervene. And this is where we can feel the hold that the link between truth and knowledge endowed with the power to judge has over us. The link that is "objectively established" cannot coexist with these other interpretations, because it bears within it a judgment regarding their subjective character. As always what resurfaces again is the opposition that allows those who "believe" that the Earth is not indifferent, who "believe" it wasn't by chance that the tsunami struck where it did on that day, who do not differentiate between its "objective reality" and the subjective suffering that it provoked, all to be shoved into the same sack, all to be defined in one go.

It is "our" word—intentionality—that allows those practices whose vocation *is not* "learning something new about what makes a tsunami possible," to be put in the same sack, the ensemble of practices, that is, whose primordial value *is not* addressing the Earth, capable of cataclysms, in a mode such that its response may be recognized as testifying independently of the intentions of those who question it. And it can do so because the question of intentionality is not excluded from practices that allow a tsunami to be explained "objectively." It even plays a crucial role here. Plate tectonics owes its "reality" to proofs that have involved competent specialists, whose questions presuppose a clear separation between *our intentions*, which lead us to put to the test and to prove; and the Earth, in the role of a witness that is reliable and thus *indifferent to the intentions* of those who interrogate it.

It is here that Virgin Mary can start to come on stage, not in great detail, because I don't have a particular knowledge on this subject, but because I owe to Elisabeth Claverie[14] the possibility of having understood that what is at stake with regard not to the Virgin Mary herself but the pilgrimages journeying to one of the places in which she appears offers an interesting contrast with what I am familiar with regarding sciences. To follow pilgrims on a pilgrimage poses the same problem, not to insult them, as following experimenters in their controversies. But the contrast is obvious: the vocation of a pilgrimage is not to "learn something new about the Virgin Mary." In both cases it's a matter of refusing to give a generality—belief—the power to explain. Far from the caricature of "naïve faith," the pilgrims described by Claverie are on a journey, but they are also engaging with a problem, incessantly oscillating between trust and skepticism, in a mode that imposes an acceptance of Virgin Mary as an actor "in her own right" to be properly deployed. It is about her that the pilgrims talk and recount what she has already "done" for them and for others. It is the possible encounter with her, at the end of their journey, that has the power to make them think and feel in a mode that is irreducible to any generality, whether critical or religious. Their pilgrimage is intentional, for sure, but its success depends on an encounter with a being who is everything but indifferent to their hope.

It goes without saying that the reality of beings associated with so-called nonscientific practices does not resist the requirements that we associate with scientifically reliable witnesses, those that permit the affirmation that neutrinos "really do exist," for example. Who will be surprised if, as we have seen, these requirements are already poisonous when they define the manner in which behavioral sciences address their "subjects"? But these same requirements simply become "stupid" when they are addressed to beings that we have dismissed as superstition, or, more politely, as representing the power that fiction or symbols have over humans.

That the Virgin Mary, as required for and involved in the "proper narration" of a pilgrimage, cannot resist the objections that experimental beings have to overcome (any more than could Mesmer's fluid) is about as significant as the lack of resistance of the most delicate, most graceful porcelain to the shock of the hammer, which the anvil, for its part, does resist. Who, then, had the unfortunate idea of putting anvil and porcelain on the same plane, that of resistance to shock? Who had the idea of constructing a plane defined by what matters to the experimental sciences and of dismissing as purely human "beliefs" whatever does not satisfy the experimental criteria of "really existing"? At the end of this journey we have passed from wonder regarding the divergent proliferation of relations to perplexity, a perplexity that no longer bears on those exotic "others" but on ourselves and the stupidity of which we are capable. And this perplexity doesn't have much to do with the question of "objectivity" in the scientific sense. Rather it communicates with the question of the repercussions of the Galilean fable of origins, that is to say, also, with the milieus on which it had an impact and which were party to, indeed an essential part of, the event, "the invention of modern sciences."

The Stupidity of Judges

What has happened to us? How did the argument about objective reality, certainly going well beyond the most excessive ambitions of Galileo, acquire such a hold that it has become a weapon for judgments that are at the same time the most routine, the most implacable . . . and the most stupid? One cannot ask this question without trembling, so much has it served to trace out the epic journey of the human species, which, for better or for worse, finds its truth with us. How much ink has been spilled on the disenchantment of the world or the eternal return of the temptation to "regress," to rediscover the enchanted world of the childhoods of humanity! To escape from these stereotypical points of view, in which words and syntax think for us, the threads of multiple stories must be woven.[15] But here, with the term *bêtise*, i.e., "stupidity,"[16] which I borrow from Gilles Deleuze, I choose to address directly something on which I wish to confer the power of scandal.

Human stupidity is not used by Deleuze to designate a general human trait, and above all, it has nothing to do with animality. Animals are anything but stupid (*bêtes*). Stupidity is a question that, for Deleuze, thinkers (Flaubert, Nietzsche) started to ask in the nineteenth century, whereas previously it was error, then illusion, that had questioned the image of thought. What they were faced with was a world in which denouncing error and especially illusion, the beliefs of "others," had become routine. What they call stupidity thus has nothing to do with the credulity that "those who know" accuse people of. Rather, it derives from the means mobilized to "protect people from their own credulity." It affects "those who know" but does so in a mode such that they dismember problems and kill questions. There is a nastiness proper to stupidity, an enjoyment that is all the more unaware of itself for being decked out, most frequently, in good intentions.

Today this stupidity is everywhere. Thus, at the start of this book, I spoke of the convention that now defines a medicine by the clinical trial that it must resist. But I did not emphasize enough just how common it is to use improvements "attributed to a placebo" as a way to explain so-called miracle cures. The placebo effect, a statistical measure, becomes an explanation that has the generality of a judgment, a cognitive judgment, dominated by resemblance, not a judgment in the sense that can make judges hesitate. Into the same, catchall, sack, are stuffed a blind statistical correlation (they took the placebo; they benefited from an improvement) and a cure that is presented not as "correlated" but as "caused," since in this case, the being responsible for the effect (a miracle worker, the Virgin Mary . . .) can be identified.

But this judgment can itself communicate with another, which unites doctors and theologians, animated by the same question, around the same apparently miraculous case: behind what appears miraculous, is there a process that can be reduced to the placebo? What this signifies is the question: Does the cure have the power to constrain doctors to admit their perplexity and to authorize the theologians to proceed and eventually recognize an intervention that is supernatural, that cannot be explained by the categories we judge to be "natural"?

In this instance, the placebo is thus supposed to form part of "nature," part of what *it must be possible* to explain naturally but which is susceptible to feeding human illusions, the risk of error with regard to the attribution of a curative power to a molecule, or the false claims of charlatans. It is thus loaded with a very great power, that of all the polemical oppositions that it arms, but it is endowed with a minimum of reality, because its mode of existence is that of a label signaling "phenomenon that for the time being has not been explained in a properly scientific mode/is to be used for polemical purposes only/ beware of anyone who uses this differently." That is to say, be wary not just of charlatans but also of anyone who refers to this "unexplained phenomenon" so as to call into question the categories through which we define "nature."[17]

The complicity between doctors and theologians is the complicity of modern stupidity, and the opposition between nature and the supernatural that they share has nothing to do with the adventure of the sciences and everything to do with the manner in which "science" has been captured by a network of polemical oppositions organized around "power": power of human illusions; power of reason or of the truth that dissipates illusions; power as the sign of the supernatural, to do what "nature" is incapable of doing; power of "nature" to operate by obscure paths that favor human beliefs in the supernatural; power of imposters; power of science that unmasks imposters; power of language to produce sectarian idioms; power of public language and argumentation to resist any drift to the sectarian; and so forth.

One cannot struggle against a network of polemical oppositions of this kind by critique alone, because, left to itself, critique might well make us fall into the relativist trap: to undo all oppositions, by denying all powers, it concludes that reality is what we name such. This is why it matters not to unlink practices and power but to link them differently. The obligations of experimenters make a reality matter insofar as it might have the power to bring them into agreement. What those who engage with the trajectory of pilgrimage make matter is the possibility of an encounter with something that has the power to transform their relationship to themselves, to their suffering, to their lives. What the reality of the neutrino and that of Virgin Mary have in common is that the practices that address these realities require them—that is to say, these practices would lose their meaning if the neutrino or Virgin Mary were reduced to human categories. In both cases—neither the physicist nor the pilgrim denies this—what we habitually attribute to humans as humans is also required. What they affirm, what needs to be heard, is that what matters for them; what makes them exist as physicists and pilgrims is neither agreement among humans nor relief from suffering in a general sense. What matters is an event which it is up to them to cultivate the possibility of.

It's not a matter here of denouncing but of allowing oneself to be marked by the stupidity of judgments that refuse to hear or understand. Whether it is a matter of "nature," of the "supernatural," of "science," of "belief," none of the terms that intervene in such judgments have the power to make one think or hesitate but rather have the power that Deleuze and Guattari accord to order-words:[18] they transform the discourse of the "victors" into self-evidence, into what goes without saying, and they can be only the object of reminders (calls to order). For those who, in our polemical history, haven't stopped fighting what they ridicule as popular superstitions, what they disqualify as untested knowledges, "nature" is not in the least bit the nature that makes practitioners hesitate and sets the experimental sciences on an adventure. It is that of the *testatores*. It speaks in terms of the "possibility of explaining" without being concerned by the difference, crucial for experimenters, between what is effectively explained and what "it should be possible to explain, one day." As for the supernatural, its definition is also empty, as the supernatural is supposed responsible for something that can have no possibility of explanation in the terms associated with what is called nature. In other words, the natural/supernatural opposition can be summed up as making rivals of two types of causality that have no other content than this rivalry.

Bertrand Méheust has shown very clearly the practical consequences of this operation with regard to mesmerized clairvoyants of the nineteenth century.[19] In cases where the hypothesis of trickery could no longer be maintained, the question focused on the possibility of reducing their "clairvoyance" to a case of hyperesthesia, manifesting "natural" faculties but developed in an extravagant manner: veritable *lupinambules* (a compound formed from *Arsène Lupin* and *somnambule*). But the possibility of this extravagant development didn't put to work those who had suggested it, or even interest them in the slightest. The verdict signified, "Move on; there's nothing to see here," and no one doing experiments in "normal" psychology took such a possibility into account. It was usually forgotten, once the disappearance of the clairvoyant allowed it to be shelved in the store of useless accessories.

The bifurcation between what truly exists, what is truly entitled to explain, on the one hand, and what can be undone, dismembered, reduced to something else, on the other, is a fearsome weapon, but there's nothing modern about it. It is, in any case, as old as the grand missionary enterprise and its destruction of idols and fetishes. But missionaries setting out to convert pagans were not "stupid" in Deleuze's sense; they were at war and knew they were. They had enemies and could be beaten. Perhaps stupidity effectively establishing itself in the nineteenth century marks the moment when the so-called modern world no longer acknowledges any enemy, only responsibility for the work of pacification.[20] No more enemies, no more fearsome powers, only those who are wandering and who need to be guided down the right path, that of the progress of a humanity detaching itself from its superstitious attachments. "We all know now" that the Virgin Mary can be reduced to human subjectivity; the only question is that of knowing which of our human sciences will establish itself as most qualified to do the work.

Unfortunately, modern stupidity, the judgment that isn't obligated by anything, that isn't frightened by anything, that can only be slowed down by the contingent obstacle of obstinate human error and illusion, is not of the order of fabulation. It is everywhere, from the ethologists who demand "proof" that animals are capable of experience (apparently, "we" alone enjoy, suffer, or hesitate) to the post-Darwinian selectionists who alone possess a finally disinterested truth about the selfish interests that determine human behavior. It passes via the full range of immediate identifications of "science" with the disqualification of what "people believe," and it involves the whole gamut of those who identify "reality" with that which has the power to resist the tests it is put to, and judgment with a right to do so.

The passage through modern stupidity was necessary in order that the "we" who figure in the question, "What happened to us?" lose any claim to represent a humanity whose radical solitude and dereliction—in a universe become the object of knowledge—it would have been "our" privilege and burden to discover. It was necessary so as to situate—truth of the relative—the scope of the ecology of practices. This proposition would not have had any meaning if it had been addressed to a missionary fighting the devil and idolatry, for example. It is addressed to practices infected by stupidity, rendered vulnerable by the order-word that demands they present themselves as "purely human."

However, a diagnosis is not sufficient to replace an order-word. One must continue to fabulate, to stimulate the appetite for thinking that alone can break the sad monotony of the murmuring deep inside us that all these histories are merely metaphors or psychosocial prostheses, because in truth—it tells us—"We are alone in the world."

CHAPTER SEVEN

WE ARE NOT ALONE
IN THE WORLD

Dismay

I learned to think about the efficacy of the affirmation "We are not alone in the world" through contact with Tobie Nathan and his team at the Centre Georges Devereux at Paris-8.[1] What's more, this book could have been called *The Djinn and the Neutrino* if I hadn't chosen to avoid anything that might be construed as exoticism. I can understand, or have the feeling of understanding, what pilgrims expect from Virgin Mary, what it is they are speaking of in terms of receiving grace: a transformation of one's relationship to oneself and one's life "under the gaze of the Virgin"—the gaze of she who understands, of she who has also suffered, and suffered as a mother suffers, not with a properly unique and unimaginable suffering like that of Christ. Psychologists can also claim to understand this "grace," albeit with words that are impoverished, without efficacy, amnesiac with regard to a heritage that "science" supposedly transcends. But before addressing pilgrimage, I would like to pause on this question of efficacy, on the choice of this term, the vocation of which is to signal that the affirmation "We are not alone in the world" is functioning as an "anti-order-word."

For us, who have dispatched the Virgin Mary and all other beings of this kind to the store for superstitious accessories, daring to affirm that "We are not alone in the world" creates mental turmoil. Straightaway a strange image of the world is set out that overflows not only with the multiplicity of experimental beings that have satisfied the tests of experimentation (neutrinos, neutrons, electrons, atoms, viruses, prions, neurons, etc.), but also with the multiplicity of "invisible" beings (spirits, gods, God, goddesses, ancestors, ghosts, etc.) that the peoples of the Earth have established relations with. We are used to the first, but in the mode whereby they have received the power to chase away the second, or to dismiss as metaphorical the possibility that "ideas" themselves, mathematical beings, concepts, fictional characters, refrains, and so forth have anything to do with "beings" for those whom they make think, make feel, and make imagine.

A world into which this heteroclite crowd of beings squeezes has something frightening about it, as if we had been set the challenge of thinking something that no one has ever thought. Every people has, in effect, cultivated "its" beings without asking the "logical" question that takes hold of us: all the same, one can't really affirm that "all of them" exist! Some sorting is needed!

Thus, the supposedly logical question is not a good question. The efficacy of the affirmation "We are not alone in the world" is precisely that what it makes perceptible is not something self-evident, something that, in the manner of an order-word, could go without saying; neither is it a terrible challenge for thinking, which "we" would have to take charge of; rather, it is a test reserved for us because it is the correlate of stupidity, which is our problem. It is we who make the logical "either . . . or" prevail, by demanding that what exists either exists "in itself," independently of us, or is something that can be judged to be, and reduced to, a simple human production. Other peoples affirm that their gods must be nourished and that, otherwise, they will die. It is we who think that if a mother "truly believed" that her dead children were the victims of a sorcerer's attack orchestrated by a jealous sister-in-law, then she wouldn't stop making her pay for her crime, clawing her eyes out, tearing her to pieces. Where witchcraft is practiced, people don't kill each other in this way, even if "protecting oneself" can signify returning the spell to its sender and even if situations can get out of control and lead to a form of pathology that everyone deplores.

The alternative "either it really does exist, or it is an interpretative fiction" results in what Bruno Latour has called the antifetishist critique,[2] prone to bringing before its tribunal all those whose "illogicality" we expose with pity or fright, so as to convince them that they must choose: either you recognize that you are fabricating your fetishes yourselves, or you affirm that they are divinities; but you can't maintain both positions at once!

It's not a matter of painting an idyllic picture of an ancient wisdom that has become inaccessible to us but rather of accepting to quit the position of a judge who interrogates witnesses. Once we accept that we are not strangers to the logical "errors" we accuse others of, this isn't so difficult.

"Recounting well" the adventures of those whom we call "creators," as if we knew what we mean, generates the same undecidability—who creates whom, the creator or the work?[3]

The very fact of speaking of *efficacy*, an old term associated with the theology of the sacraments, is part of the test of giving up logic's deadly weapon: either . . . or. In effect, the question of the efficacy of the sacraments—which, it will be recalled, divided Protestants and Catholics—has become the very example of what has been destroyed, dismembered by two rival causalities. For some, swallowing a piece of consecrated bread is an act of memory: the bread is (nothing but) a symbol; the only "real presence" is in our hearts and in our souls; all the rest is superstition. For others—rather rarely today, even among Catholics—the sacrament relates to some supernatural causality: it "really does" become the body of Christ. In both cases, there is no question; there are only answers, and general ones at that. Because both "symbolic efficacy" and divine power are held to be capable, by right, of explaining their effects. And thus also of arming a judgment with regard to the ensemble of practices that complicate this relationship of rivalrous causalities or, rather equivalently, of responsibilities.

To employ the word *efficacy* is to attempt to go back to a point before the causalities were made into rivals that dismembered the sacrament. Obviously it's not a matter of returning to the mysteries of Christian faith but of respecting its assemblage. Because it is indeed an assemblage, bringing about a relationship between heterogeneities, that the word *efficacy* designates in a generic and never a general manner. This assemblage is destroyed the moment the power to explain it happens to be attributed to one of its terms. Assemblages do not respond to a right, that is to say, to an instance that would explain and found the relation but rather, refer to the order of the event. *Efficacy* is not explained by the connection; it is the event of this connecting.

The term *efficacy* also has the great interest of communicating with something that, for us, is opposed to the event: the "efficacy" that we associate with the "it works" of technique. And that is another thing that I have learned at the Centre Devereux: the possibility of liberating what we call "techniques" from their enslavement to the reign of the "purely human" realm of means and ends. Only modern techniques present themselves in this way, as deriving from knowledges that are scientific or have a scientific appearance and that explain their efficacy on the basis of the definition of what they address. The "it works" of technique would thus testify to the value of the sciences from which these techniques supposedly derive, as transforming neutral and disinterested knowledge into means for our ends. Hence the frightening idea that if it was necessary to take "nonmodern techniques" seriously, it would be a matter of producing a neutral and disinterested knowledge with regard to the reality of beings in droves, endowed with heteroclite modes of existence that would "explain" the efficacy of the assemblages that bring them into relationship with humans.

Let us note, in passing, that in the case of the efficacy of modern techniques, too, their "scientific cause" can be challenged by a rival causality. The "technoscientific" reading of the sciences effectively places successful experimentation under the sign of an "it works" of a technical type. The whole point of knowledge that calls itself scientific would thus be for it to leave the research laboratory where it is born, because it pertains to technoscience to make its "it works" paramount everywhere that this is possible. According to the interpreters of technoscience, this reading is either a matter for denunciation or the accomplishing of an irreversible emancipation that breaks the taboos and limits that humanity in its cradle had respected. Emphasizing that the technoscientific counterproposition is restricted to the same domain as that of so-called modern techniques is pointless: obviously the Virgin Mary or spirits or ancestors are doomed to disappear, and with them, every other childish illusion of humanity. The separation of techniques that count from those that testify to the credulity of humans is similar in both cases, whether or not the sciences are recognized to have a "value" that transcends technique.

According to the idea of practice that I am defending here, whichever way causes are assigned or responsibilities are attributed, whether it is scientific success that explains the "it works" of technique or the other way round, this kind of attribution insults scientists, just as much as it does technicians, because it ignores their divergent obligations.

Tobie Nathan's proposition, according to which healers should be recognized as technicians—those with a knowledge of technical procedures the efficacy of which can be associated with what we call a cure—resists the devastating ecological consequences associated with the definition of so-called modern techniques. If in order finally to deserve the name, a technique has to derive from a valid objective knowledge, the techniques of healers, who enter into relationships with djinns or divinities, would have to be understood on the basis of what they put to work without knowing it—placebo effect, symbolic efficacy, performative power of language, suggestion, transference, and so forth. But if, as Nathan proposes, technical practice comes first, if its efficacy is liberated from its enslavement to what claims to explain it, the "finally modern" definition of what we call a psychic trouble starts to ring terribly hollow, because its "causes"—psychoanalytic, psychiatric, neurophysiological—are made primarily for explaining, much more than they are for making a practitioner think, perceive, and operate.

I would not have been capable of risking the transition from practices to the beings that these practices require, beings that practitioners know how to enter into relationship with, without the craft of Nathan's technician-therapists, who know how to cultivate the event constituted by a connection with invisible beings and how to fabricate objects whose efficacy "proves"—renders manifest—that we are not alone in the world. I would have remained defined by a certain stupidity, which says "all the same, one cannot" accept a world that is teeming with the disparate beings that human practices testify to. I would have forgotten that this "one" signals a position of transcendence, the anonymity of the judge or the *testator* who separates out the false gold from the authentic, or the position of parents who ask themselves if they can tolerate such and such behavior from their children. I would have forgotten that this disparate crowd of practices and their beings really do exist, that the question of knowing how to separate them out is the question that pacifiers ask. More precisely, it is a question that has the power to transform us into pacifiers, a question that therefore also has the mode of existence of a being that one must learn to respond to, which it is a matter of entering into relation with. Because in going unrecognized, identified with the rights of rationality or progress, this being has done what beings whose insistence one does not acknowledge, with whom one does not create a relationship, do: it has devoured us, that is to say, rendered us stupid.

To fabulate an ecology of practices is to create and to assemble the words that are capable of civilizing this idea of "ours," whose power over us is manifested with every irruption of the terrible "one" that makes us the brains of humanity, with every pacifying call to order. It is a matter of attempting to accommodate this idea, because we don't have any other choice, but also of learning to protect oneself from it. That is why fabulation must acquire an "active" formulation, the efficacy of which could be to crack this "one" whose anonymous self-evidence possesses us.

What Happened to Us?

Philosophy might feel here that it is being called into question. Because, with respect to this issue, philosophy has been active since its origins. For example, in *Gorgias*, Socrates made Callicles admit that while cooking certainly works and while cooks can successfully make delicious dishes, the art of the doctor, based on rational knowledge, a knowledge of principles, is nevertheless completely different! Socrates was certainly quite ignorant in the matter of cooking, but having made Callicles admit that it was only a set of empirical recipes, he was able to trap him and in the process make him acknowledge that if politics is to avoid being reduced to simple recipes, it must, like medicine, be based on principles, in this instance a knowledge of good and evil. And so, as Bruno Latour has shown, in *Pandora's Hope*, the stage is set: Callicles is "torpedoed," rendered completely dumb by Socrates, who hammers home the demonstration of the supremacy of the philosopher possessing this knowledge.

To be sure this is merely a "ruse," one that is likely to convince only those whom it doesn't make laugh or cry, but one could also see here the starting point for what will be more fully developed when "science" is substituted for the arts of the doctor and the philosopher, but always with the same foil, "techniques" like cooking, which are merely empirical recipes, incapable of accounting rationally for what they do. Their incredible diversity is then subject to a blind alternative: either to become "scientific," that is, to receive from a science the foundation that alone can explain why they "work," or to be denounced as the shrewd manipulation of human credulity.

However, denouncing Socrates-Plato's bid for power is not a good active formula, because it has not been fabricated but found "ready-made." To repeat something that is not only ready-made but also has been fabricated so as to render dumb doesn't have the power to escape from the operation of propagating the "either . . . or" that makes "us" missionaries and judges. To stick with the scene that Plato staged would be to forget another lesson that I have learned from the Centre Georges Devereux: the inseparability of diagnosis, prognosis, and treatment. Diagnosis is not what treatment follows on from but what "commits" it, what makes the particular therapeutic space that the treatment will actualize exist. Returning to Plato is a bit like using the DSM (*Diagnostic and Statistical Manual of Mental Disorders*) of today's psychiatrists, a sort of herbarium that allows a specific trouble to be "recognized": ah yes, what I'm dealing with here "resembles" what is being described there. But once the diagnosis is formulated, the treatment, for its part, resembles the treatments for everything else: some pills and a bit of psychotherapy . . .

An active formulation has to trouble order-words, not repeat them. It is all the more urgent to recall that the montage that creates the alternative between Socrates and Callicles, between reason and power, actually seems to announce the Galilean operation. The power of seduction that Callicles vaunts opens up the sack in which are to be found all the powers of fiction against which Galileo asserts the right of scientific proof to an exceptional status. Going back to Plato's mise-en-scène suggests a crushing continuity, such that the "we" it defines can perhaps repent—it's fashionable—but is deprived of any resources for fabulating.

Virgin Mary cannot be of any help here, because I can't think that, as the suffering mother of Christ, she would understand anything about these histories of rivalry, nor anything of the mud I evoked in the first chapter, which rises and will submerge stupefied rivals who confront each other in hopeless combat. In *Capitalist Sorcery*, Philippe Pignarre and I risked an active formula with regard to this mud itself: capitalism as a system of sorcery, profiting from our vulnerability, profiting from the fact that we assimilate sorcery to a belief, and that we thus haven't seen fit to learn how to protect ourselves from it. But this formula belongs to the construction of a political problem, and its active character corresponds to a pragmatics of struggle. Here it's a matter not of addressing the mud directly but of fracturing the normality that this mud profits from, a normality that has been so anonymous that "we" can be rephrased as "one," making common sense, polemical tradition, and the pride of situating oneself at the end of the path that human history has taken communicate directly.

Here, too, it is thanks to others that I have learned to formulate the question of this "we" in a mode that breaks any anonymous continuity. And what if we were to learn to feel that the "smoke from the burned witches still hangs in our nostrils"? This is what those who have called themselves "neo-pagan witches"[4] are trying to do, in order to activate the memory of modernizing eradication that not only destroyed witches but also peasant communities. A memory like this complicates the image of the epoch that we call the "Renaissance," as well as the "we" that stems from it. To put it very brutally: do we have to say that "we" burned the witches, after having demonized them, even if this means pitying them today, as plain old victims? Or is it a matter of stopping thinking of ourselves as archculpable or archheroic, of trying to think that such murderous stupidity is not "ours"? And in this case, doesn't the smoke in our nostrils perhaps speak not of us but of our lethally unhealthy milieu?

It has to be remembered here that this is a matter of fabulation, not of history. To try to think with the active formula "The smoke from the burned witches still hangs in our nostrils" is not to stage a historical truth, not to transform the threat of the pyre into a historical causality that it would be a matter of demonstrating through documents and archives. Nevertheless, the proposition is not arbitrary. Its value is in its efficacy, in what it is apt to make perceptible, and in the generalities it avoids "clinging" on to. In other words, it has the truth of the relative, the truth of the consequences that require it.

For the philosopher that I am, inheriting from witches insofar as they were persecuted and burned doesn't signify adoring what has been burned, taking a stand "for" witches. Nor—because above all it's necessary not to go too quickly—does it signify placing philosophy in the "satanic" lineage associated with witches, by affirming, for example, that it behooves philosophy to shake up all certainties, to subvert the beliefs that found common social life. Simply, it's a matter of recognizing that we have managed to survive in a world that burned its witches. "We" then cracks: it is neither the amnesiac "we" of the victorious nor the "we" of those who announce that if the pyres still existed, they would be the first to be condemned. It is a "we" that knows one thing: that what it has inherited is not a "great crime" the guilt for which must make us buckle but a network of order-words, all the "we now know very wells" that make us stupid, that have allowed this crime to be dismissed as case closed, and that prevent us from feeling the unhealthy pestilence that continues to poison us.

We do not know who the witches were; we have only the writings of their torturers; and it is not a question of entertaining a romantic vision of lost secrets. And still less of claiming that nothing is ever actually destroyed, that everything is sublated (conservé-surmonté), as Hegel would say, with his confidence in the ruse of reason to ensure that nothing is lost.[5] To think on the basis of the fact of destruction, the fact that we live in a world born in terror, with the devastation of a mostly rural culture, is to be able to affirm that we do not know what made historical witchcraft exist, but also to refuse the power to define witches as if this power was "normal," as if our words hadn't already served thousands of times to imprison them in our categories, to transform them into unfortunate victims, for example, with nothing very particular about them, testifying only to the barbarous credulity of their executioners. From this point of view, the "human" sciences are directly in question, because their categories prolong a past that has associated humanism and eradication. Witches were not only hunted down because of the devil, with whom they were supposedly in league; they have also been "eliminated" by those who afterward commented on and derided the "ridiculous" fear that witches provoked: they were just women, like all other women, the victims of obscure superstitions. The fact that they did or didn't themselves share these superstitions, that they did or didn't themselves "believe" they were servants of the devil, is then only a matter of splitting hairs.

Here, too, what is in question is the primacy given to the "knowledge" that confers on words an eradicating power such that nothing can grow back where what was destroyed once existed. Reducing witches to victims, calling obscurantism into question, or having recourse to generalizing sociological or psychological explanations—beliefs, habitus, ideology, mode of social perception, projection, suggestion, and so forth—is to transform destruction into a case classified as "cold," into a terrain for rival modes of explanation. These explanations converge on one point: the destruction that the witches were victims of has nothing to do with what they were, with what they made exist, but with only what they "represented" for their persecutors, perhaps with only the manner in which they "represented themselves" in a ballet that linked them complicitly with their persecutors. The operation might invoke the inverse of the Chinese adage, according to which "when the sage points at the moon, the madman looks at the finger." In this case, of course, it should be understood that the moon pointed to has no other reality than the finger that points at it.

A parallel can be made with the manner in which specialists, in saying "we know very well," deal today with those who have stopped playing the role that is assigned to the public when it is a question of techno-scientific innovation, that of satisfied beneficiary. For such specialists, it can be only a matter of problems in perception, faulty information, defective communication. The public "must be reconciled to science." That a disturbing innovation might itself be in question is thus excluded: what has to be interpreted and rectified is only the manner in which it is perceived. The public is a finger that can point only at imaginary moons: those who, in the past, believed that witches were dangerous, those who are afraid of GMOs today.

However, opting to "display one's divergence," to present today's practitioners as belonging to practices that have survived eradication, as I have been proposing, is not only to bring into question the operation that has subjected practitioners to a regime of conditional tolerance, authorizing their survival in mendacious forms. The price that is paid for a more or less prosperous survival is not just a mendacious façade. Because the obligations not cultivated by practitioners, who learn "on the ground" what makes them practitioners, without having learned the words that allow them to name and protect it from the order-words of public language, leave these practitioners themselves exposed to stupidity.

For its part, the choice made by those who call themselves "neopagan witches" today demands that the knowledge of eradication be inherited, that the "smell of the burned witches" still hang in their nostrils. They have exposed themselves to the becoming to which this name obliges. They have not limited themselves to reactivating a memory but are also experimentally, pragmatically, engaged in an apprenticeship in the present of the experience of the smoke. They indeed learn together the craft they call *magic*, a term that has been anesthetized or vilified. One may, innocuously, talk about the magic of words, of a landscape or of a moment, without further thought, as a mere cliché, but magic as a practical craft has become synonymous with the art of the conjuror, or with that of the charlatan who exploits public credulity—even if conjurors happen to unite with skeptics so as to unmask charlatans.

For witches it is not a matter of simply reappropriating what they call "magic," as if the broken lineage could be prolonged in the authentic mode that "historical truth" would demand. Rather it is much more a matter of "repairing," of remaking a connection, which obliges them to a double reparation, of themselves and of a besmirched, demonized, psychologized heritage. It is a matter of what they call *reclaiming*, at one and the same time claiming, healing, and making themselves capable of cultivating again against words that judge, normalize, and dismember.[6]

This word *reclaim* is also appropriate for what the ecology of practices demands from practitioners, for learning what to perceive "the smoke from the burned witches still hangs in our nostrils" signifies. But this formula has not yet produced all its consequences. The "we" I have associated with the smoke of the burned witches has not only been "poisoned," in the sense of weakened, separated from its resources; it is so also in the sense that it has learned to despise these resources.

The Poisons of Purity

In order to think with what we have learned to despise, I have chosen the Virgin Mary, not djinns or the witches' goddess, because pilgrims are not people who belong to "another culture,"[7] nor are they experimenters seeking to relink the past and the present.

As such, they probably don't smell the smoke of the burned witches, as they do not attract the attention of today's defenders of public order. They are survivors who subsist in the margins of a public space that identifies their practices with a kind of weakness. Pilgrims are survivors who are ridiculed as much by Voltairean minds as by those believers who tolerate them, but in the sense of "if that's what they need . . . ," as blemishes on the ideal of a faith that is "pure," disencumbered of its rituals, mediators, ex-votos, revered statues, votive candles, and processions: God speaking to man's heart in an encounter, the first meaning of which should be to "reflect" on one's faith. From this point of view, the power of transformation associated by pilgrims with the Virgin Mary shares something with hypnosis, which is denounced by psychoanalysts for being attached to symptoms, refusing the grand and terrible adventure of exploring the unconscious, the only adventure that is worthy of the human as such, perpetually in tension and never likely to be cured.

In order not to overdramatize the risks implied by evoking the burned witches and the misplaced heroism this evocation could inspire, I will try now to make felt the corrosive importance of the cold disdain that is aimed at those who "still" cling to resources that "we know" are illusory. A different lineage might be suggested here than that of the descendants of Plato, who opposed knowledge on the basis of principles to cooking recipes, or indeed, than that of the witch-hunters, who persecuted subversive powers accused of undermining social and political order: the lineage of those who could be called "athletes of spirituality," those who, since the Stoics, attempt what they themselves define as quasi-impossible, inconceivable. It is not a matter of accusing the Stoics but of underlining the catastrophe that the normative becoming of this athleticism constitutes. As Foucault has pointed out, in *The Hermeneutics of the Subject*, that a Stoic might have undertaken to make himself capable of embracing his beloved son, while keeping present to mind that this son could be dead the next day or even within the hour, might cause astonishment, perplexity, admiration, or fright. Today, however, spiritual athleticism has become normative, the instrument of judgment. We should all accept that we are "alone" in an indifferent world:

those who ask this world to sustain them are "escapists," flying away from this existential solitude. Psychoanalysts, for instance, routinely denounce all other practices addressing the troubles of the soul as so many traps, enslaving souls rather than bringing them to their incurable truth. Even epistemologists, like Gaston Bachelard, oppose the capacity to say "no" to all realist jouissance, to the "yes" that enchains knowledge to the interests of life.

In all cases, the reference to the "pure," to what transcends the boggy commons, crushes the space of practices, the space inhabited by idioms that resist public normativity. Athletes of the pure do not speak any idiom. They are situated by a withdrawal from public space. But normativity makes them eloquent when denouncing active troublemakers like false idols, idiomatic imaginaries, everything that can positively divert the human (or the subject) from the quasi-ineffable impossibility that is his vocation. Man is he who can say "no" to the crutches that make life possible.

To talk of practices is not to denounce the athletes of the impossible but to recall that athleticism is one practice among many others. The pure that designates athletic purification, the exercise of detachment that particularizes it, is transformed into a poison if it authorizes athletes to judge that it is in the vocation of being human to detach themselves from whatever their own practice obliges them to detach from. The skeptic and the athlete of the pure would understand each other with regard to the "impure" character of experimenters' realism. Similarly, the skeptic and the athlete of the pure will easily agree to denounce as "conditioning" the manner in which the pilgrims prepare themselves for their possible "encounter" with the Virgin Mary.

When Elisabeth Claverie describes pilgrims en route to a place where Virgin Mary might make herself manifest, what she is trying to approach is the efficacy of this preparation, the manner in which, during their journey together and through each other, they live a transformation of the frequently painful history that gives a meaning, for each of them, to the call that they are responding to. In this instance, pilgrims—whom Voltairean ironists would define as the superstitious par excellence—cultivate a definition of the success of the pilgrimage that resists the opposition between the natural and the supernatural. Success is not a "miracle," which is something that should be able to occur independently of any "preparation," but nor is it "getting better" in the general psychological sense. The idiom of pilgrims, who attribute to Virgin Mary the transformation that makes the pilgrimage successful—"it is she who . . ."—cannot be separated from the "pilgrimage" assemblage and its efficacy. Any putting to the test that would try to separate the success of the pilgrimage from its attribution to Virgin Mary would destroy this assemblage. As a cause, then, the Virgin Mary cannot assume the role of responsibility for the success of an experiment, the role of the guarantor that turns a successful transformation into a reliable testimony that she "really does exist."

It has to be emphasized that the divergence between the Virgin Mary and the neutrino doesn't bear on the necessity of the pilgrims' "preparing themselves." This is all the more crucial, given that to qualify this preparation as "conditioning" encourages its assimilation to an "artifact," something that experimenters worry about, and which haunts "sciences of behavior" (in George Devereux's sense). Conditioning, role play (the subject assumes the role implicitly assigned to her by the experimental protocol), performative effect of language: all these general categories arm *testatores*, authorizing them to shelve the Virgin Mary as cause in the store for useless accessories. But it is not the same for a preparation: ask experimenters if their facts don't also require a preparation, and some might stop laughing and even abandon the camp of the *testatores*, who call for facts that at the same time can be reproduced and don't demand preparation, only that they have been purified of the parasites that trouble their intelligibility. A demand of this sort is worse than a lie: it's an insult to experimentation.

Reproducibility constitutes the force of experimenters and, no mistaking, but it is what makes them think and work, what makes them learn to "prepare," and not what they might be able to obtain by simple purification. Reproducibility in the experimental sense doesn't have much to do with what, by contrast, is called repeatability: every time the head of a human or other vertebrate is chopped off, the human or vertebrate dies. In their beginnings, experimental facts are generally not "repeatable," but the aim of experimenters is not to make them such. What they are undertaking, in studying what their reproducibility requires, is a process that associates, inseparably, stabilization and innovation. To stabilize a procedure is to learn what it demands in order to become robust and intelligible. It is not "repeating" that this involves but the exploration of variants, putting consequences to the test, creating filters which can be shown to eliminate effects that are indeed "parasitic." In short, learning to prepare.

Although repeatability is often presented as what defines "facts" that are susceptible of science, one could say that the *production of reproducibility*, the production of conditions that are suitable, or adequate, for a successful relationship, is the major achievement of the experimental sciences. This production doesn't designate the event that makes scientists dance,

but its laborious aftermath, the obtaining of a double conjoined definition: the nonhuman now has a stabilized definition, which corresponds to
well-defined practical operations. Such a success can be celebrated in the
manner of the animal invention of lift, when the new role of air was stabilized and the new, aerial, milieu was inhabited. The animals this concerned
didn't "repeat" but explored along divergent paths what "flying" might signify for them, in a double process of innovation and stabilization of what is
"suitable" to the type of flight adopted. Each type of "discovering how to fly"
is the event of a new composition that confers a cascade of significations on
what has been "enrolled," thanks to which a still indefinite set of histories
will henceforth be invented, in new modes.

It is because such a process of stabilized enrollment is in itself an
achievement that experimenters are indulgent toward "discoverers" who
are retroactively convicted by detectors of fraud of having "altered" their
results a bit. They know the difference between a deliberate fraud, aiming
to deceive colleagues, and a discrete selection of the "best" results, leaving
aside those that are judged "aberrant." What matters is that the risky bet
has been won, that the experimental procedure has become reproducible,
that techniques have been perfected, and that, from now on, the result is
susceptible of being obtained by no matter what competent colleague. This
doesn't exclude "sleights of hand," "artfulness," a few informally transmitted
"tricks," but it does imply that if a beginner doesn't succeed, it is the beginner, not the standardized procedure, that will be called into question. And
the major achievement is the re-articulation of the know-how into a reliable instrument that can be used without special competence.

However, the parallel between the preparations of experimentation and
that of pilgrimages to the Virgin Mary comes to a halt very quickly, because
the ultimate success of experimentation is indeed instrumentation, when
what comes to the rendezvous becomes a reliable part a "public" equipment. Similarly, the success of birds is to be able to "count on" the air, as we
(sometimes mistakenly) count on our legs to stay upright. In both cases,
the success is routine, and the Virgin Mary cannot enter into such routines. Who would dream of "counting on the Virgin Mary" in the way one
counts on electricity when switching on a light? The conditions that are
suitable, and also the modes of composition, diverge.

Despite this divergence, which we will address shortly, what may have
become perceptible is the poison of purity, the judgment made concerning
those who still have need of "prostheses" to enter into a relationship with
what they appeal to. Against athletes of purity, who, in the name of the

ineffable relationship to transcendence, are contemptuous of the "mummery," the devotion and the rituals of pilgrims, it can be maintained that far from being a prosthesis, something one ought to be able to do without, the pilgrims' mode of preparation translates an apprenticeship, the cultivating of a connection that is, at the same time, a knowledge production.

Pilgrims are not stupid. They would be the first to recognize that what responds to their hope—she whose possible presence at the end of their journey has the power to make them think and feel in a mode that they associate with grace—has to be distinguished from the stories that they tell with regard to this encounter. But the same distinction also concerns neutrinos. When physicists say, "Henceforth, neutrinos have a mass," what they say can certainly be understood in the mode, "There are, in the world, neutrinos endowed with a mass," but what they are celebrating is more the fact that in order to successfully establish a coherent experimental relation with what they call "neutrinos," their apparatuses for doing this had to take into account what, for them, is designated by a particle with mass. In both cases there is indeed a production of knowledge, a knowledge immanent to the relationship.

The question of what the being that a relationship has been established with might be "in itself" has no practical sense. However, what matters, what "describing well" depends on, is that what is characterized in terms of the practical role that has been cultivated, the successful relation, not be identified with a "purely human" convention. The success of the relation is not a right but is of the order of the event involving a being that might have not responded. And the knowledge that is produced speaks of the being *insofar as it responded*.

Against the celebrated "we only know what we make," it can be affirmed here that the "we only know what we learn to fabricate a relation with," and that we know it according to the categories that correspond to the success of this relation, its robust stabilization, in the case of the experimental relation, the preparation required by the specific link, bearing the signature of the Virgin Mary, between suffering and grace, in the case of the pilgrims.

To Domesticate?

Fabulating here imposes a certain prudence, because giving the impression of "having understood everything" would be a failure: the production of a truth that is valid for all, which everyone would have to consent to. I do not speak in the place of practitioners. My question is addressed to the milieu whose ecological transformation I am fabulating, which nowadays poisons practitioners by inciting them to present themselves in a mendacious and/ or stereotypical manner.

Associating causes, insofar as they obligate, with the question of connecting with beings as heterogeneous, nonhuman, has sense only if the association opens up the appetite for, creates the possibility of, escaping the unavoidable dilemmas that make us stupid—does that "truly exist?"/ "is that merely a human construct?" Abstaining from reducing practical causes to a metaphor, from bringing everything that matters down to a subjective or social selection, is a test whose difficulty is strictly coextensive with the hold that the "normal" power of judgment has over us, the habit of thinking, perceiving, and feeling "with the smoke of the burned witches still hanging in our nostrils."

It is from this point of view, which has the truth of the relative, that one can go a bit further and give a renewed consistency to the proposition according to which we know only what we can fabricate and only insofar as we know how to fabricate it. This proposition, which is usually associated with Giambattista Vico, is most frequently mobilized through the example of the watchmaker, who knows his watch because he made it, adjusting it piece by piece. In this example, the pieces are not conceived as beings but as the means to achieve what the watchmaker wants to achieve. By contrast, one might say that the operation my proposition designates is of the order of "making an alliance." That we know only to the extent that we have succeeded in "making an alliance" could be the formula for a practical culture affirming that "we are not alone in the world." The operation of "making an alliance" in effect implies success in connecting with something that is not us, something with which it has been possible to create a relation but which will never become ours, which will instead oblige us to learn what the connection requires.

As an example that may be good to think with, one can evoke something humans experimented with long before becoming watchmakers: the taming or domestication of animals, the set of practices that enabled humans to "make an alliance" with animals. Of course, the term *domestication* as an example of practical culture will enrage: one will hear in it "domestic," a "being" placed in our service, something that might be appropriate for electricity or even for the problems posed by nuclear fusion (researchers are still trying find how to "contain" it . . .), but certainly not for the Virgin Mary or for what makes a musician or a philosopher create. Concepts make one think along the zigzagging lines of a witch on a broomstick, and it's the philosopher's responsibility not to get carried away (enjoying enraging) or to block the movement. In this instance, one might recall that "domestication" derives from *domus*, home, a place inhabited by humans, and that its pejorative meaning is linked to the idea that we know what a human is, that the human *domus* possesses stable categories to which domestication subjects animals. And yet "domesticated" electricity was not subjected to preexisting categories but transformed ways of life, habitats. The notion of domestication becomes much more interesting if it is related to the creation of a suitable *domus*. What *domus* is suitable for parrots or for primates, if it is a matter of learning what they can become capable of?

Pilgrims are not seeking to learn with regard to the Virgin Mary, for sure, and they know that the suitable *domus* is nothing other than themselves, something that will make them apt, in Claverie's terms, to pass "from a closed economy of destiny to an open economy of salvation."[8] The emphasis is placed here on the becoming of the *domus*, not on what can be attributed to it, that is to say, not on the extension of its "properties." The Virgin Mary escapes all appropriation, and in her case, the term *domestication* is doubtless not suitable. In *A Thousand Plateaus*, Deleuze and Guattari created the concept of the refrain, the efficacy of which is to "to assure indirect interactions between elements devoid of so-called natural affinity, and thereby to form organized masses."[9] But they distinguished between territorial refrains and deterritorialized refrains, which transform the first from the inside, opening them up to "cosmic" forces that cannot be subjected to forms, that it is a matter of making perceptible but which, becoming perceptible, carry human territories off in a cosmic becoming. One doesn't domesticate the cosmos. The stake for practices here might be related to needed protections, for "many dangers crop up: black holes, closures, paralysis of the finger and auditory hallucinations; Schumann's madness, cosmic force gone *bad*, a note that pursues you, a sound that transfixes you."[10]

Domestication is thus only suitable for certain operations, those that can, in one mode or another, be associated with a production of knowledge, those that can be associated with a "property." But even in this case, the terms of the relation are incapable of explaining the connection. Perhaps the first of the operations of "connecting," the most familiar, the one we tend to consider "normal," is called "comprehending." But who hasn't entreated someone else to "pay attention" when he doesn't comprehend something that nevertheless seems perfectly comprehensible! It is indeed a matter of entreating because no one can tell someone else what she has to do in order to pay attention. It's only later, once the attentiveness has allowed something new to exist, has made possible a new connection, that what it was worth paying attention to finds its reasons.

The formulation for the entreaty itself might be as follows: "Accept being modified by what you are dealing with!"; "Accept suspending the perceptual or cognitive routines that make you think you know what you perceive or what these words are saying!"; "Accept that without which what you are dealing with would have no efficacy." This entreaty, to "accept," may be efficacious, but not in a mode susceptible of explanation. More precisely, explanations are relative to the stakes of transmission. The mathematician is happy that the student has at last read the statement of the problem, which is supposed to contain everything there is to comprehend. Physicists, as Thomas Kuhn has clearly shown, are not satisfied by comprehending the statement: they demand the production of a territory. This is the role played by exercises in which the student learns to "recognize" the manner in which what he has comprehended permits the problem to be posed and resolved, whereas the mode of "paying attention" proper to scientific paradigms—territorial refrains—is stabilized. But elsewhere, this "learning to pay attention" can be interpreted as testifying to a weakness: look how easily influenced she is; she started to like that oeuvre only after others told her why they liked it; she is copying everyone else!

There is nothing really very surprising about the fact that so-called scientific psychology teaches us very little about what "learning to pay attention" can signify, and it's difficult to see the fact that lack of attention is now classified as a "trouble" (attention deficit disorder) as a triumph of scientific reason. The pragmatics of attention, the different manners of concentrating, of freeing one's mind, of being watchful, or even of letting one's mind wander, by contrast, belong to spiritual techniques that are much older than psychology, techniques which, contrary to psychology, take seriously the fact that an efficacious encounter with what they make matter can be in play here.

To talk of experimental practices in terms of "domestication," the creation of the *domus* that will make the relation that experimenters have successfully obtained reproducible, has nothing pejorative about it. Any resemblance between theoretico-experimental categories and the all-terrain notions that allow for something to be defined without having entered into a stabilized relationship is effectively broken. These all-terrain notions, which are dear both to *testatores* and to sciences like psychology that have "carved out" a territory for themselves by driving out the anecdote in favor of the repeatable,[11] and by accepting the statistical approach, which demands repeatability, as synonymous with science, designate what can be called antipractices. Any possibility of learning is thus excluded: because it requires the indifference of both whomever it samples and whatever is supposed to become manifest in statistical correlation; the statistical approach by definition forbids itself any practical culture with regard to the *domus* that would allow for accommodation and stabilization, that is, for any learning to establish a relation.

The order-word "a fact must be reproducible" always lies, but it doesn't prevent experimenters actively creating reproducibility, nor pilgrims cultivating the art of the event which they name "encountering the Virgin," nor animal tamers, breeders, or ethologists learning what makes an animal capable of learning. This lie can, however, weaken some practices by subjecting them to the temptation to present themselves "badly," indeed even to mistreat the great achievement that their successes testify to. The nineteenth-century "clairvoyant" subjects whom the "magnetic relationship" brought into existence knew something about this when they accepted, even demanded, tests that would have allowed them to "prove" that their gifts testified, in a reliable manner, to a "force" that was foreign to what can be attributed to the human. In fact, the tradition of magnetism constitutes a unique case of hybridization between therapeutic practices, "spiritual" practices aiming to open up the *domus* to something that can never become a "property," and practices of domestication aiming at proof. It's a rather sad history, as is testified by the case of the unfortunate Alexis Didier, who spent his life guessing the answers to uninteresting questions in an attempt to "prove."[12] This was a desperate task, because a proof requires an interested milieu, not the hostile milieu that Alexis Didier had to confront, in which each skeptic that he finally convinced was treated as gullible by the others. Like trying to empty the ocean, one spoonful at a time.

This is why the surrealist poet André Breton's proposition, related by Bertrand Méheust,[13] interests me, even though it may disappoint those for whom the only real question is one of knowing whether clairvoyance "truly exists." For Breton, it was necessary to free the procedures associated with the production of the magnetized's "clairvoyant" faculties from the hands of the adepts of proof, as well as from those of the physicians, so as to cultivate them in a more favorable milieu, one that would not impose inept tests on them: the artistic milieu. In other words, cultivating the refrains that produce clairvoyance as recipes, the preparation for a transformation that allows the seer to establish a relation with something that cannot be domesticated.

What Breton proposed—and put to work with his group—was not "the" solution but an ecologically pertinent proposition. That artistic practices might welcome, without trying to explain, something that the magnetized, the martyrs of proof, wanted to prove the existence of, doesn't mean that art is an "irrational" practice. Quite the contrary, each practice, situated by the type of relation that it is a matter for that practice of making exist, of cultivating, stabilizing, can be called "rational" in the sense that *ratio* signifies "relationship," and that the cultivating of relationship is also a creation of knowledge. But what differs is what this cultivating aims at, in the sense that it is a practical—that is, a collective and transmissible culture. Will the type of events that it conditions permit an extension of the *domus*, as is the case with practices that produce "positive knowledge" or domestication? Or will this culture instead create the *domus* that is appropriate for events on which one will never be able to "count," encounters that escape domestication? Or is it a matter of yet other alliances, alliances that could be placed under the sign of the taming of what nonetheless remains wild, that make for a "philosophical tradition," for example, or that fabricate healers, or that link judges to the question of "justness" . . . ?

Generic Propositions

No generality can be maintained when the question of knowledge (is it valid? anecdotal? objective? rational?) is replaced by that of productions of existence. However, one can hypothesize that *generic* terms would be useful, terms that, contrary to general definitions, stimulate an appetite to distinguish, terms whose efficacy is to bring practices together on a plane that no transcendence hangs over. A plane that cannot bring them together without having them exhibit their divergence.

Beings, *cause*, and *efficacy* might belong among these terms, as might *force* as Deleuze and Guattari use it. But one should not start to complain that this is fraudulent, the misappropriation of a "scientifically well-defined" term. Not only is the term evidently much older than it was in its scientific acceptance; also, its (mis)appropriation doesn't borrow the authority of physics any more than the (mis)appropriation of efficacy is inscribed in a theological problematic. This (mis)appropriation leads instead to situating physics differently, to "denormalizing" the thought-event that the Newtonian force of attraction constituted.

Everyone knows that Newtonian force was a scandal until physicists agreed to think it starting from its effect, acceleration. Besides being foreign to Newton's thinking, in which force expressed the power of God at work in the world, this definitional articulation between a force as cause and its effect is instead nothing other than the *domus* that allowed for force to be accommodated as a "property" but also transformed physics, for which what henceforth took primacy wasn't matter (become mass), space, or time, but what assembles them, the *measurable equality* of cause and effect.[14]

However, it took a long time for the adventurous version of rationality associated with Newton's audacity to be forgotten. The chemist Gabriel Venel, author of the article "Chymie" in Diderot and d'Alembert's *Encyclopédie*, heralded a science that, like Newton's, knew that "nature brings about the majority of its effects by means that are unknown, that we cannot count its resources and that it would be truly ridiculous to want to limit it, by reducing it to a certain number of principles of action and means of operating." And the same confidence resonated among nineteenth-century magnetizers and twentieth-century parapsychologists: if they could just succeed in producing solid facts, facts that would impose themselves despite being inexplicable according to the accepted norms of explanation, their colleagues would be obliged to accept what they otherwise judged to be impossible.

In order to give to force its generic sense, it's this confidence in the example of Newton's audacity unifying in one fell swoop the explanation to be given to seemingly disparate phenomena that it's a matter of breaking with here. The difference pointed to by Venel between "chemical bond" and "physical aggregate" was finally domesticated but only by the bizarre means of quantum physics, without a "force" proper to chemistry. As far as the general explanation evoked by those who study so-called psy phenomena is concerned, the idea that their hour will come signifies that there will come a time when it will finally have been possible to domesticate them in the mode that experimentation demands. In other words, not only in a mode that is reproducible but in a mode that makes just one question matter: can "psy" effects be mobilized and staged in such a way that they can make the difference between the different manners of identifying what they would then be the reliable witnesses of?

I hope that the reader will have been struck by the bad taste of this requirement. We don't know what "psy" phenomena, premonitory dreams, for example, testify to. What we know about is the manner in which a life can be marked by these kinds of dreams, when they occur "in the raw," outside of a culture of the dream, a culture that knows how to decipher a message in and through the dream. And those who know cannot be experimenters, because their concern is to decipher, not to invent the apparatus that would allow the dream to be transformed into the reliable witness to whatever it is that would explain them.

In short, as long as a reference to physicist's forces is active, force is not a generic notion but a misleading one. The divergence that matters bears on the role that different practices would confer on a "force" insofar as it is called on to make a difference. Does this difference require a culture of reception? Is it liable to explicate *itself* but not *be* explained as the effect of something more general? Will those who experience the efficacy of "forces" be liable to be transformed into apparatuses that allow others, competent colleagues, to convoke the "same" forces in order to enroll them in an experimental proof?

Forces pose the question both of the apparatuses that allow them to manifest themselves and be recognized by what they produce, and of the *domus* that could perhaps confer on this production its signification. Unlike the traditional philosophical distinction between *natura naturans*, naturing nature, nature that produces for and by itself, and *natura naturata*, nature represented, identified, and explicated, there's no need here to mourn the loss of a "reality" that would remain foreign to the knowledge produced about it, no resigned acceptance faced with the limits of "our" knowledge. Certainly the physicist "represents" physical forces, but his success is precisely that what is represented is not subjected to the categories of this representation but responds to them. No loss to be mourned here but a very particular practical success that is not susceptible as such of defining general categories or a limit. The equality of cause and effect, the *domus* of forces that physicists characterize as "central," is anything but a generic category: it has practical meaning only in the physics inherited from Galileo and from Newton.[15]

The practice of experimentation can be called an attempt to *convoke* a being—one that could be given the name "neutrino," for example—in a mode that makes of it a reliable witness. But pilgrims do not convoke the Virgin Mary, although they do evoke and invoke her. It's interesting that everyday language has resources that are able to nourish generic distinctions without which the plane it is a matter of creating, or more precisely of making felt, would crush what it should bring together. And that it does so, notably, with the group of verbs whose common root implies a form of *appeal* or *call*: evoke, invoke, convoke, provoke, even revoke.

In a characteristic manner, each call implies a distinct mode of coming to the call. To *invoke* is to appeal to a being as endowed with power, whereas what is *evoked* can transform experience simply from the fact of its becoming present: one evokes a memory, but one invokes the law. To *convoke* a being implies a call that one can expect to be answered by a being with which a fragment of a story can be entwined, while to *revoke* a being is to exclude it from stories that are to be intertwined. Finally, to *provoke* implies that in order to intervene, what is called must be "activated"; it must pass from inaction to action.

One could say that each use of verbs that make an "appeal" or a "call" implies a "force." An appeal is made to a being, but each mode of appealing to indicates how this being must be appealed to, which assemblage it is called on to participate in, for what efficacy, with the truth of what *domus*. Among the modern practices that make use of "appeals," the law is one of the most explicit, and if there ever was a fable, it is that of a self-sufficient legal statement, a law that it would be enough to invoke so as to be able to judge well-established facts or explain the judge's decision. A law only speaks, only has force, if it is convoked, and it will speak in the terms of each party when pleading its case. As for the facts and witnesses, the manner in which they are convoked by each party and the manner in which what they are the witness of is invoked or revoked by the judgment are enough to indicate that it is not they who speak. The decision belongs to the judge, and it is the judge's qualities of hesitation, not capacities for deduction, which will be honored if need be.

The beings appealed to by practical cultures that we judge to be superstitious also explain themselves through the roles that they play in making a difference that matters for practitioners. Rather than speaking in terms of properties, which is appropriate when the vocation of the appeal is a connection that should become reproducible, one can speak instead of *ethos*, because what responds to the call does not, in so doing, stop being a *matter of concern* that demands attention and vigilance. Here, what corresponds to forces is a whole ethology defining the intentions of beings, what they require, how best to enter into commerce with them, to protect oneself from them, to fool them, even.[16]

It would be interesting to explore the manner in which every practical culture gives a meaning to the verbs *convoke, invoke, evoke, provoke, revoke*. Contrary to the binary classifications dear to ethnologists, putting the relevance of the distinctions that these verbs make to the test would not involve a combinatory system that gives power to "either . . . or" disjunctions, but rather a composition.

Hence placing experimental practice under the sign of successful convocation above all permits the designation of a singularity, which points toward the "realism" of experimenters, toward what might make them dance. But this convocation often implies a provocation, and it is difficult to talk of experimental practices without also bringing on stage the evocative power of models or paradigmatic cases. As for the invocation of progress, which transforms the event of the successful experiment into the putting to work of a right to conquest, transforming the experimenter into a predator, and the revocation of so-called philosophical questions, for example, which is the ritual accompaniment of the presentation of a scientific field, one of the bets of this essay is that experimental scientific practices have no need of them. But these two correlated components are definitely part of what many scientists judge to be essential, as much for the force of successful experimentation as for the confidence that ought to inhabit a true scientist.

For other practices, invocation, the appeal to a being invited as a power, is crucial, but here, too, the mixture matters. Healers invoke and convoke when what we call healing requires negotiation and commerce with the being that has intruded into a life, but for their part, pilgrims invoke and evoke the sufferings of the Virgin Mary, for example, and the success of the pilgrimage, the "grace that is received," isn't the miracle manifesting the power of the Virgin Mary, but the manifestation of her presence, of her gaze that will transform their own relationship to their suffering. As for the efficacy of philosophical concepts, no doubt they bring a dimension of provocation to the fore, but such concepts make thinking communicate with a deliberate operation of putting at risk. On the philosopher's part, such an operation implies revocation, the deliberate distancing of what proposes a ready-made solution—in the present case, for example, sociological or psychological categories that eradicate. But this doesn't happen without the convocation of something that for each philosopher will act as a safeguard against the intoxication of omnipotence, constraining her through a relation of fabrication with her concepts. Nor does it happen without evocation, without the "thinking in the presence of" that makes thinking exist as resistance.

Making the list any longer would be to sink into the intoxication of omnipotence and risk failure; it would be ventriloquism that makes one speak for, in the place of, and that systematizes without having experimented with the means for distinguishing between verbal violence and creation. The sole aim of linking practices to the divergent culturing of manners of "appealing" was to make perceptible something that already inhabits common experience—albeit in a mode that is vulnerable to metaphorical or psychological reduction. This approach must protect itself against the temptation to commentate, subjecting what one hasn't experienced the practice of to what one believes one knows about it.

Nothing of what I have written transcends the ambition to confer an efficacy on the affirmation "We are not alone in the world." It was a matter of making one think at the point where the fabulation borne by oxygen having become a vital resource and air having become an integral part of the phenomenon of lift were interrupted, with the bird celebrating sunrise, testifying thereby to the "force" to which it is sensible and that makes it sing. In the first two cases, one could conclude that the history of living beings created an unanticipated bringing into relation, for sure, but one that belongs to the "reality" that the sciences know how to understand and which, once understood, is prolonged by techno-industrial development—those that enable planes to fly, for example. With the bird singing the rising of the sun, a tipping point was reached where, suddenly, the bird alone was responsible for what was then no longer a "relation." We can now say that the Sun, such as it makes the bird sing, can certainly not be convoked in the laboratory or by modeling or through astronomer's observations. It is that whose provocative force has poetry, painting, singing, or prayer as its witnesses. It has the truth of the relative, as indeed does the Sun that can be convoked by scientists.

The reader who discovers that she is no longer shocked by this statement, that she may have become the *domus* capable of entertaining it, will then be able to ask herself the perplexed question: But how are we to understand each other if neither reality nor reason can bring us into agreement? That is the question that the ecology of practices asks.

ECOLOGY OF PRACTICES

Questions of Civility

Some years ago, during a public talk organized at Chateauvallon, a celebrated biologist spoke to us about the origins of man and our hominid ancestors who ventured into the savannah and experienced the advantages of bipedalism. What I heard was a fine story, and, like the rest of the public, I was charmed. But during questions, a hand—that of Tobie Nathan—was raised. He politely pointed out to the biologist that while he was well within his rights to evoke his own anonymous ancestors wandering in the savannah, he could not, on the other hand, speak for Nathan and his ancestors, who did not live in the savannah, but in Cairo, and who were not anonymous, because he knew their names. Nathan went on to explain to the flabbergasted biologist, who was oscillating between perplexity, amusement, and indignation, why the ancestors of a people form a part of the questions that no one has the right to deal with in place of that people, short of wanting to destroy them. The public had just had the privilege of being present at an unfortunately rare event. One practitioner had told another practitioner that his manner of speaking was a declaration of war, in this instance because it implied and announced the destruction of peoples who know who their ancestors are and who see their thinking being dismissed as fiction, in the name of a scientific narrative that, for its part, claims validity for anyone whomsoever.

Apparently, the public, much like the speaker, heard nothing more than a paradoxical remark without any thinkable consequence. This happened before the "science wars," which otherwise it could have been an episode of: Nathan could have found himself being accused of irrational and obscurantist relativism and assimilated to the defenders of creationism. The public talk at Chateauvallon occurred before September 11, 2001, but had it not, he could have found himself in a still more dangerous situation, being accused of encouraging murderous fanaticism. Everything that sets the bad example of "resistance to scientific rationality," whether it is a matter of monotheistic religions or the customs of people who know who their ancestors are, is susceptible today of being put in the same bag. As for me, taken aback like everyone else at first, it got me thinking on the basis of an aspect of the "role" of the ancestors described by Nathan: a civilized people presents itself to others by presenting itself as descending from ancestors *who are not those of others*. Thus it presents itself as different, but in a mode that doesn't imply that it has what others might not have. Every civilized people effectively has ancestors, and their differences are not rivalrous but enable exchange. "Enable" but do not guarantee, in the Habermasian manner, for example, which links "humanity," "reason," and "communication." Exchange is not the norm; war or conflict are not something pathological. What is ensured is the *possibility of exchange*, not its normality.

Nathan's proposition implies that those who present themselves as "scientific"—that is, as rational, neutral, objective, and so forth—are not "civilized." Either one admits that in certain situations, all can display these qualities, in which case all might just as well present themselves as "humans." Or one insinuates that these qualities are foreign to those to whom one is presenting oneself. One thus defines them in their place, and exchange, for its part, is defined as impossible. What is imposed is, in this case, a pedagogic relation.

It is indeed this noncivilization of science that was directly translated by the normality with which a biologist defined the ancestors of all humans as humans. To be sure, one could say that the two uses of the term *ancestors* diverged and that even if it didn't matter to Nathan, his ancestors, in the paleoanthropological sense of the term, wandered in the African savannah like the ancestors of all other humans. But this merely translates the fearsome efficacy of the reference to the "human"—what matters to biologists and to paleoanthropologists allows them to define something that is supposed to concern someone, no matter who, whether she knows it or not, whether she likes it or not. Because "ancestors," even in the paleoanthropological sense, cannot be assimilated to a trait that would be of interest only to specialists, like the average mass of the human brain, for example, or the mode of articulation of the jawbone. What is being constructed is a story loaded with signification, that of the "origins of humanity."

There are some real conflicts that now occur on this issue, around the skeletal remains of the famous "Kennewick Man," for example. American Indians who wanted to get one of "their" ancestors back and bury him according to custom, and researchers who claimed him for the purpose of study—considering him all the more precious because he might testify to a people more ancient still than those from whom the American Indians descend—went before the judges. Remarkably enough, the judges demanded that the researchers "present themselves" as the American Indians had done, so as to make it possible to decide between the interests of two comparable parties. This scandalized the researchers: Did they not serve the universal, that which ought to concern whoever wants to escape from archaic particularities?

The American Indians were finally beaten on appeal. On July 6, 2005, the researchers were at last able to gather with their instruments at the Burke Museum in Seattle, where the Kennewick Man had been resting, inaccessible to everyone since 1998. He has become their ancestor and is too interesting for the date of the "reburial" demanded by the American Indians to be conceivable. He belongs to science indefinitely.[1]

Thinking this conflict in terms of the ecology of practices doesn't offer a solution, and it's not a matter here of adding to the opposing arguments made before the judges. The test associated with the ecology of practices is to accept the simple fact that *there is conflict*, that is to say, to avoid all recourse to the pacifying order-words that would make the American Indian tribespeople backward, still prisoners of their beliefs—implying that the judges had no need to hesitate. The ecology of practices excludes a position of transcendence that distributes reason, and the judges' decision doesn't in the least imply such transcendence. On the contrary, it signifies that the mechanisms charged with regulating the conflicts between interests and verifying legitimacy have been set in motion. The decision was based on the lack of proof for the link the American Indian tribespeople claimed with the Kennewick Man. But what translates the unhealthy character of the situation is that, having won, the researchers profited from their victory in the manner of predators who have rediscovered their self-assurance: those who slowed them down momentarily were nothing but "prey," defined by beliefs that it behooves science to transcend. It appears that the episode didn't make them think, didn't teach them the words that might have avoided humiliating those who were defeated.

In fact, for the majority of scientists, there was no need for the judges to "hesitate," the very necessity of their intervention being of the order of a pathology that translates a rising tide of irrationalist relativism. By contrast, betting on the coexistence of divergent practices doesn't bear primarily on solving conflicts. This bet presupposes the irreducibility of conflicts to a deviation from the norm, and on the part of the protagonists, it requires that the question of civility, of what the possibility of relations obliges them to, become a question that matters as such, a question that is capable of making for hesitation.

Knowing how to present oneself is part of this question of civility. To fabulate an ecology of practices is to fabulate practitioners who know how to present themselves in a mode which is that of their *force*, in both senses of the term: in the sense that force designates that with which they enter into relation, learning the *domus* it requires, but also in the sense that it designates the manner in which their obligations bind them, makes them diverge from general matters of concern. In other words, a "civilized" ecology of practices would oblige all practitioners, scientific or not, to learn how to separate the force that made them into practitioners from every generality that would allow them to situate and therefore judge others.

Needless to say, such generalities are legion, and avoiding them demands an effort that some will judge artificial. And this is where the proposition of an ecology of practices encounters a first objection. Is not this learning to "become civilized" a commitment to an "ecological correction" that would force practitioners to "behave" under the watchful eyes of those they might shock?

This is an objection that matters to me, because if the ecology of practices was to come down to submitting to rules that ensure the possibility of coexisting, the proposition would be reduced to a law of "polite silence" without the slightest interest, because it could be addressed only to the good will of those whom it concerns. Why expect scientific practitioners, who benefit from all the shortcuts in thinking that references to "science" and "progress" allow, to give up these privileges, except by crediting them with a privileged relationship to the true and the good? How can one expect them to feel obliged by questions that their education has given them the habit of judging to be secondary and even irrational? By means of what miracle would they agree to complicate their relations with the protagonists of state and industry who profit from these habitual judgments?

To respond to this objection is to recall that the proposition of an ecology of practices is not atemporal, claiming a truth that "should have" been laid down from the very birth of what has been called "modern science." It belongs to our epoch. And it does so not just because where

once pacification was given free rein, active, recalcitrant groups are beginning to appear, who make perceptible the manner in which routines of
judgment and legitimacy insult them. It does so also because the shortcuts in thinking that define "science" and "progress" are being turned back
against researchers and rendering them doubly vulnerable: in relation to
their traditional allies who now impose their own priorities upon them
and in relation to a "public" that is beginning to break with the role of satisfied beneficiary that had been assigned to it.

Let us recall the manner in which biologists, in the affair of GMO
crops, were caught between a rock and hard place: on the one hand, their
mobilization by industry, and on the other, the bringing to light of the
profound ignorance, the irresponsibility, of those among their number
who had announced the making of incontestable and inevitable progress.
To accept that they are mobilized by industry, as most of those interested
do, is not only to accept money but now also to accept new constraints, notably the channeling of research by the question of patents—patents to
apply for or patents already in place. As I have already emphasized, it is
to accept the possibility of the destruction of what binds and obliges them:
each researcher will be bound by the role that she plays in the knowledge
economy, which will demand secrecy and condemn hesitations and un-
welcome controversies as susceptible to weakening a promise of renewed
competitiveness, the only thing that will henceforth prevail.[2] Can we imagine biologists who would gratefully recognize that anti-GMO protests enabled them to ask questions that would probably have been overlooked or
for which they wouldn't have found funding? In other words, biologists
who would feel indebted to the protests, thanks to which they can become
better—more demanding and lucid—researchers?

The ecology of practices acquires its signification on the basis of the
transformation of relations that is announcing itself, and it affirms the dis-
symmetry between those recalcitrant groups who are learning to meddle
with what was supposed not to concern them and the traditional allies of
researchers. The latter now threaten researchers with destruction, whereas
the former demand that researchers learn to fabricate new types of links
with them. As I have already underlined, it happens that one species can
disappear from a habitat when others have changed their behavior for one
reason or another. Or that the species survives at the cost of a change in its
behavior, which can signify *living on the margins* in precarious dependency,
but can equally signify *living differently*, thanks to different relations with
others. The proposition of an ecology of practices does not appeal to the

good will of scientists. What it fabulates, what it seeks to make an appetite for exist, is the difference between undergoing and resisting, between maintaining one's habits despite their no longer really being successful and daring to think where shortcuts in thinking and fables that justify them have previously reigned.

It cannot be emphasized strongly enough that a proposition of this kind does not have the power to bring about what it appeals to. Its interest is that what it requires is also what is required for a future in which scientists might escape from submission. What the new knowledge economy, like the predator-prey ecology, demands is scientists who are "not civilized," who are able to abstain politely from judging, perhaps, but incapable of separating themselves from the generalities and shortcuts of thinking that enable the "nonscientific" to be judged, identified with the "irrational." A good predator has to be deaf to what would complicate its relationship to what it defines as prey. As for the capacity to enter into relationship with groups whose learning trajectory translates their becoming-practitioners, obligated to learn by the questions they learn to meddle with, this demands researchers who are able to present what matters for them in a civilized manner. Without reference to the "scientific mind" that legitimizes the radical asymmetry between the lucidity they collectively cultivate with regard to their discipline and the opportunism and/or naïveté and arrogance they feel is legitimate toward everything else.

Ecological Disasters

But all the same, someone might say to me, your fabulation dictates that you maintain a polite silence about the fact that as a philosopher, it is surely impossible for you to accept anything and everything. Take clairvoyance, for example: can you accept that someone is capable of foreseeing that in three days you will encounter a handsome, dark-haired man with green eyes? Isn't it necessary to set down some limits all the same? But already it is possible for me to retort that forcing me up against the wall in this brutal way, staging the alternative between "believing in" and "not believing in," is precisely what practical cultures, where clairvoyance is not accompanied by any kind of sensational claim, avoid. For instance, it is said that in the learning trajectory of Buddhist initiation, it is "normal" to "see," but that making a big deal out of it, experiencing/experimenting with this trajectory to attain this goal and, worse still, transforming clairvoyance into a "fact" that authorizes claims to knowledge, is a trap. Thus it is not about maintaining a polite silence but about learning to pay attention, including—moreover—to the manner in which the philosopher has just been addressed: as someone who has to make a ruling on what is entitled to be a component of the world, on compositions that are possible and those that are impossible.

But here, too, it's a matter of stimulating the appetite that makes the difference between a fabulated possibility and the utopia of good will. Is abstaining from normative judgment such a bad thing? Have normative judgments helped in thinking the composition of our worlds? I am not among those who are frightened by people consulting tarot card readers and astrologists; I even think that whatever their theories might be, these practitioners cultivate practical knowledges that might well be more pertinent for helping the people who consult them than the majority of degree-holding psychologists. However, and without even saying anything about the power of expert economists, those pseudoclairvoyants, *I really am frightened* by the irresponsible stupidity with which we unthinkingly and offhandedly give admittance to the beings that our techniques allow us to convoke: in a mode that constitutes a real recipe for ecological disaster.

And yet denunciations and normative judgments have not been lacking in this regard. What hasn't been said about a world that is "dominated" by technology, by a mastery that is blind to "human values"! Who hasn't held forth on technology out of control, which should be at the service of man, not enslaving him! Who hasn't fired back, waxing lyrical about the greatness of human history, the breaking of all its taboos, the necessity to defeat the static values that hold it back, and the continuation of what is already at work in the evolution of living beings, the invention of the means for making limits explode, for overcoming obstacles, for creating new, hitherto inconceivable possibilities!

Perhaps one can now understand the sense and interest of the speculative detour that has made me multiply the "beings" whose reality is required for thinking the creation of relationships as the event of the connecting of the heterogeneous, irreducible as such to any generality. The idea that we ought to be able to "judge" such beings in the name of general norms is rather inappropriate. It's a matter of knowing how to accommodate them, because they are apt to enslave those who believe themselves able to define them as simple "means" in the service of an "end," as in the service of human needs, for example.

ECOLOGY OF PRACTICES

It will perhaps be said that this kind of definition has already been ripped to shreds by critique. Sociologists and historians have, for example, shown that technico-industrial innovations create, rather than satisfy, "human needs." My point is different. It's not a matter of critiquing but of learning to be frightened by the worrying naïveté of the very idea of assimilating techniques to "means." The issue is not that the disarticulating of techniques into the (scientific) principles that would permit their mode of action to be defined, and the "ends" in relation to which this action would be a "means" would be erroneous, without correspondence to the "reality of things." It is rather that it designates our "reality," which eradicates the collective concern that should be cultivated: how to accommodate what techniques make exist among us.

Of course, only "modern" techniques can be presented on the basis of principles that are scientific or that have the scientific allure that would justify their efficacy. We will not deny the successful symbioses between scientific research and technical innovation, but we will emphasize that it is not just "ancient" techniques, those that could not be presented in the "modern" manner, that have been disqualified, but also what was often cultivated with them, the necessity of "paying attention," of learning to negotiate, to accommodate, and to protect—in short, to think. The recent "discovery" of the eminently unsustainable character of technico-industrial development here reminds us that symbioses that are successful from the abstract point of view of performance are not necessarily successful from the point of view of the art of paying attention. Because (scientific) principles are supposed valid whatever the terrain, they have a real efficacy here: they permit the passage from "one terrain to another" without the necessity of what might be called the taking up of a relay, the creation on *each* terrain of the *domus* that a new technique should call for in order not to be reduced to a "new and better manner of satisfying the same need." In this sense, the definition of modern techniques as "the application of scientific principles" is a sure recipe for a nonculture of technical innovation.

I've already discussed GMOs. I could mention the crazily imprudent way in which the truly unique anthropological experiment that consists of exposing children to the techniques of advertising has been carried out. Or the disturbing way in which, through blind statistical correlations between genetic differences and somatic or behavioral traits, "risk groups" have multiplied, announcing new forms of discrimination that are no longer based on "prejudices" but on a "good" science that merely seeks the well-being of all. But I will take up a much more ambiguous example.

When French biologist Jacques Testart, dubbed the "scientific father" of the first baby born, in 1982, using IVF treatment, noisily, publicly, withdrew from research into medically assisted pregnancy, he made an audacious gesture that earned him some forceful criticism from his colleagues. He effectively broke the convention according to which technical progress is "in the service of humanity." He had participated in this progress, in extending to women artificial insemination techniques invented for livestock, and he had experienced the extraordinary and unexpected—that is to say, the unanticipated boom of "the same technique" (in the biological sense of the term), as it moved from the domain of animal breeding to become an additional means being offered in the medical "marketplace." He did not affirm that the techniques were "bad per se" but only that we were not capable of accommodating them, even if, to this end, we mobilized a whole troop of ethicists, psychologists, and psychoanalysts and made them judges of the authenticity of the "desire for a child" and of a couple's capacity to distinguish between biological and symbolic paternity.[3] According to Testart, we were playing with fire, because these techniques brought what I am calling forces into play in a mode that he reckoned was destined to unchain them and that neither the army of shrinks nor restrictive laws (for instance, those about embryo screening and selection) would manage to control.

It's not a matter here of asking whether or not I think it is "good" that, thanks to artificial insemination, lesbian couples can have children and that they can, as a consequence, claim official recognition as mothers. Or if, like Pope Benedict XVI and horrified shrinks, I'm going to call on the dogma according to which every child "needs" a biologically "true" male father and a biologically "true" female mother. The ecological disaster is signaled precisely by the poverty of the dominant alternative: either to be opposed, referring to expert diagnoses concerning the future of such children that act as real curses (the third generation will be psychotic!), or to affirm that the parents are couples "like all other" parents and—the technique being a simple means to an end—that the children will also be "just like other" children. In other words, the situation has been taken over by an order-word: the children to be born, thanks to those techniques, must be guaranteed to be "just like any other." The strange continuation of something that began in the mode of "like livestock," only more complicated.

This order-word standing in for argument is not unique to the question of modes of "artificial" conception. It's much more a signal of what we ask of our "artifices," our technico-social innovations: they have to be neutral, simple means for ends that remain unchanged; they must not obligate us to any creation, any practical culture. In other words, there is no need to think, to create, to give to "purely technical" artifice the *domus* that is able to accommodate it. We know that technical artifice will transform us, but we require that we know nothing about it.

Because things do not, generally, go "well," there is something to be learned from the manner in which, sometimes, they do, where techniques are actually accommodated as the bearers of the possibility of transforming a world. This is notably the case when, an experimental apparatus having been transformed into an instrument (something that forms part of the possible consequences of its success), the question is posed of the possible adoption of this instrument by other communities of practitioners. "Adoption" must be understood here in the sense that it is an event that matters, that questions.

To be sure, there are scientific fields wherein the prime mover seems to be instrumentation that has come from elsewhere. This is the case with neurophysiology, for example, where imaging techniques give so many responses that it's no longer necessary to create original questions, to risk hypothetical relationships, to hesitate about the obligations of a science of what is called "the brain." It's also the case when the principal goal of research programs and collaborations is to justify the acquisition of a machine, the possession of which then becomes an end in itself. In this regard, one ought to speak of a "mobilization" that has no "link" between interested colleagues, or a "knot" with the functioning brain,[4] a mobilization that produces "empirical data" that are essentially no different from specimens that explorers bring back from distant countries—even if they can now, as sometimes is the case, be patented.[5] And one ought to recognize that this type of mobilization is precisely what the new knowledge economy requires: scientists who hesitate at nothing, who will do everything it takes to stay in the race, to obtain the most powerful, the most sophisticated instrument.

However, the strength and the vitality of a scientific practice, its capacity for hesitation and creation, does stand out in the controversies, discussions, hopes, and fears that a new instrumentation possibility stimulates. The questions, "In what ways is this relevant for us?" and, "How will this transform us?" bring into play a "we" whom it behooves to hesitate regarding the manner in which it will be modified by a new instrumental possibility. Here no practitioner has the naïveté to think in terms of new means to an end that would have the privilege of remaining unchanged. No practitioner will think that it is a case of "supply" responding to their need. It goes without saying that no practitioner will think that the operation of adoption can be considered as "normal" and that a shortcut over "paying attention" can be taken.

This capacity to pose the question, "How to adopt?" is something that no normative judgment can ever be allowed to take shortcuts around. It is not unique to scientific practices: it typically happens in judiciary practices, for instance. In fact, it is the correlate of what I have defined as the obligations of a practice, what makes practitioners hesitate.

The possibility that the ecology of practices, of practitioners who learn to present themselves on the basis of their obligations, seeks to make exist does not, of course, offer any of the guarantees called for by those who affirm that one cannot do without normative judgment, that a definition of the general interest to which everyone—anyone—owes her obedience, is necessary. What matters is not the right to exclude but a capacity to "include" that demands a culture of both attention and recalcitrance.

Distinguishing between Sleepwalkers and Idiots

If, at the start of this chapter, I was dealing with modern scientific and technical practices, this is not because I suddenly forgot Virgin Mary to the benefit of the neutrino but because in our epoch and in our countries, which are so proud of their pacifying order-words, they are the practices that set the tone. This includes how one speaks about pilgrims, reduced as they are to simple private matters, to be tolerated once they do not trouble public order. In the matter of artificial reproduction and lesbian (and gay) couples demanding to be treated "the same as everyone else," hostile (French) Catholics make a great deal of noise, but they are careful not to get themselves accused of letting their "private" convictions intervene in a "public" matter. They speak in favor of the "child's interests" and mobilize expert psychologists who make a big drama about the destruction of the family, "scientifically" promoted to the status of the condition indispensable to the psychic organization of the child.[6]

In other words, the division between the public and the private traverses the practices that I am trying to think together, and this division also marks my proposition. I can search for the words that concern practices that have survived by making alliances with interests defined as public, so-called modern practices, but what the practices that modernity has triumphantly relegated to the private might become capable of is an unknown of the question. Between the two, there is no imposition of a hierarchy, but it seems vain to hope that what has been defined as private might be able to experiment with the manner of making its obligations matter, as long as so-called modern practices have not initiated a process of disengagement in relation to the public order.

So let us come back to these "modern" practices, to practices that have managed to retain a certain "vitality," despite a mendacious alliance with the public order. This term—*vitality*—which most often designates something one benefits from as "normal," which one can enjoy as of a right, "as long as you've got your health," is *unfortunately* adequate. As it happens, it makes it possible to feel the imprudence testified to by the beneficiaries of a force—who enjoy it as if it were a "normal" property, one that can be possessed. If the knowledge of peoples who do not live as though "alone in the world" is to be taken seriously, this imprudence can be formulated as follows: he who believes he "possesses a force" is possessed by it, and he will expend tremendous energy extending the hold of this force possessing him.

There's nothing exotic about a knowledge of this kind. The image of the sleepwalking practitioner corresponds to it, the practitioner who, above all else, must not be woken up when he is balanced on the edge of a roof, because if he realized what he was doing, he would hesitate, experience vertigo, and fall off. (In this image, balancing on the edge of a roof is always what the sleepwalker is doing.) The sleepwalking practitioner requires that one not ask him to hesitate where his practice doesn't obligate him to. In other words, what he requires is to extend as far as possible the difference between what matters to him and what he judges to be secondary or anecdotal, to maintain the strictest division between "his" hallucinated world and a "reality" that he must be able to ignore. Sleepwalking practitioners claim the right to preserve the hallucinated image of a practice that progresses autonomously: "Do not wake us up, because if we had to acknowledge the importance of what we judge to be secondary, nothing would protect us any longer from the cynical descent into sad relativism. We would be defenseless against the rather dreadful idea that our successes are merely conventional and that our practice is merely a function determined by interests that transcend it. Let us remain blind, and let us educate our students into the same blindness."[7]

Hence one can understand why Thomas Kuhn's proposition seduced scientists, why they recognized themselves as working "paradigmatically." Kuhn effectively linked the fecundity of scientific communities endowed with a paradigm to the fact that the latter is integrated as "normal," generating an incognizant faith. That Kuhn proposes a new public representation, linking fecundity and faith, and not fecundity and rationality, wasn't a worry for scientists: what matters is that the paradigm justifies the requirements of scientific communities with regard to their closure, with regard to the difference between what obligates them and what is of no concern to them.

The manner in which modern practitioners—like Joliot-Curie as described by Bruno Latour—quite expertly bustle about in order to gain the support and many resources his work needs evidently has nothing to do with the surefootedness of the sleepwalker oblivious to his vertiginous situation. Consequently one might say that the image of the sleepwalker is a lie pure and simple, destined to keep public (credulous, incompetent, etc.) opinion at a distance. But the sleepwalker, or the possessed, can be perfectly lucid. What identifies him or her as possessed or as a sleepwalker is the abstract character of that lucidity. Typically, sleepwalking scientists decipher the world in terms of opportunities, something that permits the extension and reinforcement of the importance of what matters to them, what possesses them, what makes them predators.

The ecology of practices bets on the difference between identity and belonging. On the one hand, the sleepwalking practitioner presents himself as the bearer of the definition of what obligates him. He will say, "In my capacity as (a physicist/lawyer/doctor, etc.), I must . . ." and will treat whatever puts this identity in danger as an enemy. Belonging, on the other hand, can be associated with the figure of the idiot,[8] one who speaks "idiomatically," but also—as re-created by Gilles Deleuze—one who resists public urgency, for whom there is *something more important.* If the sleepwalker requires that he not be woken up, that he not be forced to give his attention to the consequences of the frontier that he incessantly fabricates between the inside and the outside of his practice, the idiot, for his part, refuses to submit to the generalities that would transform what obligates him, what forces him to hesitate, into anonymous and consensual norms that can be stated in a public language.

Between the sleepwalker and the idiot, the difference seems slight, as both are strangers to the ideal of submission to public interests. The prospect of a stable peace born of goodwill is no more valid for idiots than it is for sleepwalkers. But from the point of view of the ecology of practices, this difference matters. It opens the possibility of escaping from the foil constituted by the image of inescapability associated with the predator-prey ecology, which licenses the conclusion that if there isn't anything to guarantee an understanding, then nothing but normative rules can stop the strongest from devouring the weakest. To be sure, for the idiot there is something that is more important than the understanding between humans. But what distinguishes the idiot from the sleepwalker is that she doesn't need to be blind. The manner in which she hesitates will certainly be uncommon. The hesitation of a practitioner will always be a practitioner's hesitation, not that of a human supposed capable of renouncing his particular interests in the name of the superior interests of peace. However, whereas the sleepwalker can only be a predator, because asking him to hesitate—to wake up, that is—is to want his ruin, the idiot's mode of hesitating is not reinforced by a judgment that defines what must absolutely not make her hesitate. What can make an idiot hesitate is an unknown.

Sleepwalking practitioners correspond to a predator-prey ecology, identified with a single kind of stake, on which their survival depends—capture a prey/escape from a predator. An ecology of this kind is a bad abstraction. Ecology gives us examples of successful symbiosis, or of coevolution, such as that to which flowers owe their scent and insects their specific sensitivity. But neither the flower nor the insect had to demonstrate the goodwill to overcome their differences and discover what unites them beyond their divergences. They do not have an understanding, do not pursue a common goal. Symbiosis is the creation of a local articulation between divergent interests, an articulation of heterogeneities that remain heterogeneous, that links without unifying, that composes without subjecting the terms that enter into composition with each other to a common interest.

Understanding the reference to ecology in a literal, not a metaphorical, manner is to confer on the *creation* of articulations between practices the value that such a creation takes on in the history of living beings, which is that of an event. It is the importance of an event of this kind that translates the manner in which a practical community is susceptible to hesitating regarding the possibilities of an instrument that comes from elsewhere, or the manner in which it makes count for itself what counts for others. This event needs idiots, for whom there is something more important than mutual understanding, because what *has to be created* is an articulation between idioms, an articulation that cannot be generalized into a "basic agreement" that dissipates misunderstandings and transforms divergence into a simple negotiable discrepancy. The creation of an articulation has the truth of the relative, of the situation, or of the question, which demands the coming into existence of this articulation.

CHAPTER EIGHT

No "ecological" articulation can be inscribed as such in a logic of necessary consequences; it cannot be subjected to the power of the "as you have accepted this, then you must also accept this. . . ." The flower's interests are not articulated with the interests of the insect in general. A scientific community that knows how to accommodate an instrument that has come out of the research of another community is not committed in any way by the obligations of this other community. The general power of the "therefore" is the power of public language, whereas each idiom cultivates its own "therefores" in divergent modes. The ecology of practices gives the creation of articulations between humans the value of an event, to be sure, but these are humans to whom their respective practice has given the force proper to idiots. It is, then, incapable as such of guaranteeing a general peace or even a peace that is generalizable little by little. However, what it allows for the fabulation of is a *culture of knowing how to hesitate* against the vindication of anesthesia that fabricates predators. To make articulation into an event is to ask that each protagonist know how to recognize themselves as situated by their idiom, that is to say, as incapable of putting themselves in the other's place as such and of being able to say in the other's place to what consequences they would be obligated by an articulation.

Of course, this doesn't signify abandoning the common, public language that allows for understanding, but its restriction to what in fact has been stabilized, to what does not obligate the locutors. That we are able to understand each other most of the time doesn't in the least bit guarantee that that understanding is normal, that it ought to prevail once the obstacles that hinder communication are removed. This fact merely underlines that a person is no more defined by the belonging that obligates her than the wolf is defined by its behavior as a predator. Whether she is a sleepwalker or an idiot, the practitioner is also someone who, like everyone else, has routines that are more or less the same as those of everyone else. Someone who is preoccupied by unemployment, by her children's future, by global warming. The manner in which she deals with this may well use common language, but this doesn't, in the slightest, mean it's lacking in importance for her—quite the contrary. The ecology of practices does not imply that the preoccupations of "citizens" have no role to play. What it does imply, what it does demand, is that none of the specialist idioms that participate in a matter of "public concern" be able to claim to define the manner in which the problem in question must be posed publicly. The language of public concerns must then not allow for a neutral position but must put to work an active and demanding culture, a culture that is caustic, in the sense that it would attack any amalgam, that it would force all protagonists to present themselves in terms that foreground their obligations and not by means of claims to a legitimacy that has to be imposed on everyone.

A prospect like this is political, and pursuing it would be to quit the speculative register of the ecology of practices for that of science fiction and the exploration of what might become possible if the question of the rising "mud," which is in the process of destroying all idiots, no matter whom, didn't pose itself first. The sole aim of what I have been proposing here is, I recall, to stimulate the appetite against the order-words that destine practitioners to fight against each other as this mud submerges them all. That is why I won't go any further on this theme, but will instead confront one last objection, doubtless the most serious. If practitioners cannot have a "basic" understanding as humans endowed with goodwill, the question poses itself: what really can ensure the possibility of reaching an agreement—beyond articulations that are of benefit to those between whom they are made, like that of the flower and the insect? We come back to the major complaint: one cannot count on the event; a public language is necessary, as without it we are at risk of civil war! And it's a matter now not of fabulating but of "dramatizing" this risk of war.

The Efficacy of Dramatization

To fabulate is to seek to occupy a terrain, such as that of the fable of Galilean origins, in a mode that stimulates the imagination and the appetite for possibilities that this fable denies, that it chokes off with a sigh, "Yes, I know, but all the same." All the same there is a difference between what scientists imagine and what humans imagine. . . . All the same, one cannot accept everything. . . . The operation of dramatization, for its part, is addressed to what collides with the operation of fabulation, with what puts fabulation itself at risk, and which it cannot ignore without collapsing into an unpleasant utopia.

To dramatize the threat of a civil war is not to "make a big drama" out of it, nor is it to give this threat the power to explain the history of eradication that has made us "modern." Among all the threads woven by history, the argument, "But that would lead to civil war!" has probably not played a dominant role; yet it is the thread that we are most vulnerable to, having the greatest power to bring together under the sign of the sacrifices necessary for a shared understanding. And in the case of the ecology of practices, this necessity could be stated: if what obligates is also what makes for divergence, what creates idiots who are recalcitrant with regard to the general interest and prevents understanding based on goodwill, then it would be better to sort things out "among humans." That is to say, among those welcoming generalities that will ensure communication, that is, will allow everyone to learn how to put themselves in the other's place and will legitimize the habit of disqualifying, of dismissing as private, whatever could divide.

One can, of course, counter that the understanding and the goodwill that are supposed proper to humans are an abstraction, and that if idiots have to be destroyed, they will be so primarily by something completely different from the decision to give peace precedence. It isn't in the name of peace that scientists are "called to order" today, but in the name of mobilization for an economic war from which no one must escape. However, such a response is too easy, because it is limited to silencing. Choosing to dramatize is to accept the test of the objection; in this instance, it is to recognize that the threat of war is a correlate of the ecology of practices, but without that threat having conferred on it the power to produce the "and therefore . . ." that unites for peace. It's a matter of conferring on this threat the power to oblige thinking. How is peace to be thought if the protagonists refuse to yield to its necessity, if they prefer war to the betrayal of what obligates them?

From the outset a strange contrast is suggested. Although peace as the order-word in the name of which submission can be required is a powerful argument, thinking on the basis of peace as a question makes for a rather weak position. I recall here the derision provoked by the term *irenicism*, associated with the thinking of the philosopher Leibniz, in a Europe torn apart by religious wars, that is to say by a confrontation that nothing more important transcended. Since then, irenicism has remained associated with an abstract, even hypocritical, optimism that wishes to blind itself to the hatred and the violence that rip history apart. But one must also recall the accusation that was addressed to Leibniz by some of his more perceptive contemporaries: *Herr Leibniz glaubt nichts*; Herr Leibniz believes nothing. Here it is not a case of derision but of dismay or scandal: Leibniz seems to say we are right; he approves of our theses, and he repeats them, but he does so through a strange operation that is all the more insufferable for being ungraspable; they no longer confront head-on the theses against which they were to prevail, and they lose their power to mobilize for war.

Leibniz was a philosopher but also a mathematician and a diplomat. Perhaps it is because he was a mathematician that he knew the difference between submitting to and being obligated by: mathematicians are obligated by the mathematical beings that they make exist, and they are so with such intensity that many do not hesitate to accord what obligates them their own mode of existence. They explicitly affirm that these beings are not their creation, that, as mathematicians, they do not have mastery over them, but that they are instead indebted to them because it is to them that they owe what it is that makes them mathematicians, creative not submissive. And perhaps it is because Leibniz knew this that he was able to create a conceptual system capable of transforming him into a philosopher-diplomat, never contradicting his multiple correspondents but creating a translation of what they forcefully claimed, which would—if they accepted it—make a possibility for peace exist where a logic of war prevailed.

Leibniz's exhortation *calculemus!* has nothing to do with the exhortation to submit to a public language that would impose its terms on heterogeneous idioms, in the manner of economic exchange values: how much is the right to pollute this lake "worth"? for example—it being understood that if I pay, I can. Leibniz's is a language that isn't imposed on anyone, because, as a mathematical invention, it allows what every idiom makes matter to be "saved." Its virtue is to translate and articulate what seems to be contradictory, to formulate what I have called the obligations that make idioms diverge in such a mode that they no longer require the defeat of what seemingly contradicted them.

Leibniz failed, and his failure can certainly be explained in a mode that condemns his hope: people demand submission; the truth is not the truth if it cannot legitimately demand submission. Those who draw this sensible and sad conclusion generally exclude themselves from the judgment. They themselves and an elected few, unlike "the people," have no need of submission to the order-word of public peace. Others will not condemn the unrealistic character of Leibniz's hope but the project itself. They will lay claim to the link between truth and polemic, the refusal to give way on their desire, the refusal to bend the formulation of their requirements in a mode that deprives those requirements of their power to harm. If the operation of dramatization exposes whoever attempts it, it has to learn from Leibniz's failure and attempt to inherit his *calculemus* differently, in the mode of a revival that complicates his trust in the possibility of peace. What if Leibniz hadn't protected himself well enough against the poisons of war?

Leibniz knew these poisons. The enjoyment of a polemical truth, of a truth that defines any possibility of peace as treason, is the enjoyment that he associated with damnation.[9] But he makes his appeal to "calculus" as if—once obligations were "saved," and except for the damned—a peace offering would suffice in order for weapons to be laid down. The weakness of the Leibnizian *calculemus* is that it presupposes what it would like to generate: a world in which all agree to coexist peacefully, to "bend" the formulation of their obligations to what peace requires, asking only how to do so.

In a sense that is technical, not critical, one might say that Leibniz missed his target and that he did so perhaps because he thought he could connect too directly three components of his own genius: the mathematical invention that transmutes contradictions into articulated contrasts, the zigzagging of the concept, and the twists and turns that the role of diplomat demands. In *The Fold*, Gilles Deleuze has spoken in this regard of the two floors of Baroque style. We have to imagine Leibniz running from one floor to the other. Above, there is the system turning round on itself, with doors and windows shut, and below, friendly and bewigged, Leibniz, who receives, debates, corresponds. Receiving an objection, he runs to the upper floor, inputs the objection into the system, and comes down with the means to make what his interlocutor proposes correspond to a proposition from the system. But the Baroque style doesn't suit the ecology of practices. What is situated "above" is a veritable diplomatic "headquarters," which, while certainly not a center that imposes submission, nevertheless puts philosophical reason in command. It is delocalized but one nevertheless, because the system, which creates convergences, functions everywhere that divergence provokes conflict.

The Leibnizian philosopher certainly isn't king, and he doesn't owe his role to any transcendent principle, only to the mobile intelligence detached from worldly passions that his system obligates him to. It remains the case that not only does a practice, that of the philosopher, take it upon itself to set out the plane on which divergences coexist—something that effectively arises from conceptual creation—but also that this creation claims the efficacy of a rationality that would come to be acceptable as such by all.

This detour via Leibniz, via the operation that "dramatizes" his failure, was necessary, because what is dramatized here is the temptation par excellence that threatens the ecology of practices. It is not enough to accept that peace doesn't pass through the suppression of obstacles to the general rules that ensure ready-made communication but through the invention of articulations between heterogeneities. And it's not enough to say that this articulation doesn't preexist, that it isn't general but has to be created precisely where conflict threatens. This creation also must avoid any generality; it must not communicate with the general prospect of a saving of the world to which everyone would be supposed to aspire, in which everyone would be inscribed from the moment that a diplomatic "headquarters" of the Leibnizian kind would let them do so without betraying their obligations. If not, the creation of a common reason that erases conflicts between contradictory requirements will become a "cause" that obligates all practices, transforming the articulation to be established into what matters for each. The only hesitation would then bear on the "how" of articulations that are expected and desired by all.

In mathematical terms, one might say that the Leibnizian solution corresponds to a problem that hasn't been fully unfolded. The idiots dreamed by Leibniz authorize a solution that is too simple, because they themselves have a philosophical dream, the dream of a common world in which each would have accepted the conceptual possibility created by the philosopher and would become active participants in the dynamics of its realization. Correlatively, while the articulations created may well be local, they will nevertheless be caught up in a general dynamic in which each will be celebrated unanimously as progress. Except by the "damned."

To protect oneself from this temptation, it's a matter of sticking to the constraints of an ecology that, in order to thwart the threat of civil war, cannot mobilize the slightest transcendence, even the transcendence of a peace to be invented. One cannot go from the logic of, "This means war!" to a unanimous cry, "No war, above all else!" The ecology of practices must stick to a possibility that is always local: it *may happen* that the protagonists in a potential conflict envisage this conflict as "something that could perhaps be avoided." Correlatively, the philosopher can certainly be an artisan of peace, but conceptual creation allows no shortcuts; it doesn't permit the philosopher to speak in the name of a "we" whom her concepts would in one manner or another be capable of convoking.

In the framework of an ecology of practices then, the question of peace cannot be posed in general. The possibility of avoiding a war will be envisaged by the interested parties only in the cases where one generally invokes the necessity of a public language to avoid war, to arbitrate between contradictory requirements. Furthermore, it will be envisaged only in cases where "causes" "force" the belligerents to war, where otherwise they would betray something that they define as more important than their interests, something that makes them idiots. It will not be envisaged in the case where "human" interests are in conflict, nor in those cases where the strong profit from their strength to enslave the weak—it is the law, in such cases, that is, supposedly, equipped to arbitrate. This is why to pose the question of a possible peace is always to take a risk for a potential belligerent. Obligations that are at risk of betrayal vibrate; hesitation is necessary; and the possibility of peace cannot require and presuppose a "general desire" for peace, which every possibility of war would be a welcome occasion for fulfilling. It can only require and presuppose a practical culture that actively separates a cause from the right to conquer, that doesn't honor its extension as a verification of its legitimacy.

If the Leibnizian *calculemus!* needs to have an ecological sense then, the "we" that it implies cannot be one of practitioners who desire to understand each other, who hesitate only as to the modalities of that understanding. Because the invention is the creation of an articulation, it can certainly be called a "calculation," but in the sense that calculation implies *not a reason but an artifice*, which creates both the "we" and the terms of the calculation. Like every creation of a relationship, this calculation cannot be produced in general, but only with regard to the particular stakes, the local conflict of requirements that makes war probable. And like every creation of a relationship, it cannot be conceived independently of practitioners, those who, in this instance, are effectively obligated by peace as possible. These practitioners can be called diplomats because the difference between probable war and possible peace is the diplomat's concern.

As it is a matter of practice, it is not a question here of describing the empirically existing diplomats of today but of thinking the obligations and risks of an art that is, no doubt, as old as war itself. Dressed in tailcoats, togas, or boubous, "diplomats" can recognize each other because they practice an art that has nothing rational about it, in the sense associated with argument, debating, exchange, everything that defines humans as ideally interchangeable, masters of their thinking and reasoning. An art that has nothing to do with a peace-saving mission that is honored and saluted by all, because diplomacy is rather seen as hard work and diplomats are always suspected of indifference with regard to causes that demand heroic greatness. An art that depends on an interstitial possibility, always to be created and never guaranteed, between what obligates parties and the manner in which these obligations are formulated and correlated with requirements that seem logically to mean war.

Diplomats are often reproached by righteous people for never "going to the heart of things," for remaining at the surface, that is, for being happy with superficial arrangements instead of bringing about genuine reconciliation. Their "arrangements," which have the uncertain precariousness of fictions that only just hold together and which do not confer on an agreement the power to reduce a conflicting past to a misunderstanding are held in contempt. Moreover, this is also why psychoanalysts reproach nonmodern "healers," when they talk about suggestion, the manipulation of the transference, and so forth: only psychoanalysts would go beyond "symptoms." It's also what Plato reproached the sophists for when they presented themselves as "therapists," even though their art wasn't based on any kind of science of the right and the true. In this sense, linking the question of peace with the art of diplomacy responds to the constraints of ecology. Because this art, which produces artifices, translates well the fact that those who eventually, possibly, succeed in agreeing were not "looking for an agreement." The agreement doesn't arise from misunderstandings having been overcome or from an equilibrium being returned to, through the resolving of a temporary conflict. If the agreement can indeed "hold," it is not because the world has become "simpler," rid of what perturbed it. If it happens, it does so because a world that has become more complicated has been created, sufficiently endowed with extra dimensions for what was contradictory to have a chance of coexisting.

Leibniz was a diplomat; his language was shrewd and sinuous. But his weakness was wanting to play every role at once, that is to say, to imagine a diplomatic "headquarters" in charge of everything. Leibniz's universal reason unflaggingly translates reasons for war into possibilities for peace, but it gives no meaning to diplomats failing to agree, because for him diplomats who meet each other are in fact delegates from the same headquarters. However, the diplomat as I understand her, cannot do anything simply by herself. She "represents" a cause that, in the instance of a diplomatic failure, is susceptible to obligating war, and she needs other diplomats delegated by the other parties with whom war is probable. In other words, the art of diplomacy depends on the fact that the "parties" between whom war is logical nonetheless decide to "give peace a chance" and to delegate diplomats.

However, a Leibnizian idiot in the service of peace in general might perhaps take part in the creation of actively, speculatively, neutral places that welcome diplomats. Perhaps with his presence, in the discrete mode of humor that suggests without arguing, he might participate in the imagination of possibilities that will make the diplomats hesitate. But he can never be confused with a diplomat, because he can never be accused by those he represents of betraying their cause.

Since I started having recourse to the figure of the diplomat, a "frequently asked question" has been: "Who can be a diplomat?" Or, more warily, "Is this a new name for the practice of philosophers?" It is therefore worth specifying that when I envisaged this figure for the first time,[10] it was in regard to those "social" sciences that should, John Dewey proposed, prolong the experimental logic that was already at work among living beings. At the time, I hadn't read Dewey, but for me, in any case, the problem was not posed in the same manner as it was for him. Dewey was thinking at a time when these sciences were still experiencing themselves as in search of their practical specificity, and I was thinking in the face of a situation that translated the failure of his hopes: not the invention of a continuity but the order-word of a resemblance. It is, therefore, not in continuity with the experimental sciences but on the basis of the particularity of their diverging obligations that I posed the problem of the social sciences and in particular on the basis of the manner in which proof, which obligates the experimental sciences, can, in the case of the social sciences, become a "requirement": something that destines them to take situations of "weakness" as their object, the only situations that allow for a stable differentiation between those who ask questions and those who must answer them in a complying manner, giving to the scientist the freedom to interpret and draw his conclusions.

It was not a matter for me of marking a dramatic rupture, for instance of concluding that, given the compliance such sciences require, they were not scientific practices, with the correlative consequence that specialists in the social sciences had no other choice than to despair or to grit their teeth and continue to mimic "real science." In *The Invention of Modern Science*, when I spoke of the "sciences of contemporaneity," in which the production of knowledge, for one group, and the production of existence, for the other, are strictly contemporaneous, I was already seeking to convey a divergence in obligations that I still didn't know, at that time, how to name.

Subsequently, two characteristic traits of what I have since associated with the "figure of the diplomat" started to appear. The first is that a situation that overtly differentiates between the active scientist and the complying subject can and must be avoided. The specialist in the social sciences, such as I envisaged her, has to address people who belong to groups producing their own requirements, people capable of asking their own questions, creating their own position, claiming their own point of view about the way they are addressed. What this implies—it's the second trait—is that in this case the specialist is always exposed to the accusation of betrayal. This problematic of betrayal is "practical," not psychological. It results from the fact that if the group studied by the sociologist has the required "strength," it pursues its own ends and has its own voice. The researcher then is not able to define the fact of becoming the group's "spokesperson" as a success, as is the case in the laboratory. It is rather threatening her own position as a scientist. Contrary to the physicist's propensity to dance when he has succeeded in conferring on the neutrino the power of conferring on him the role of spokesperson, the sociologist will be denounced by her colleagues if she can be suspected of having been enrolled by the group she is studying. But if she refuses to be enrolled, this group can itself accuse her of "instrumentalizing" it for the benefit of her science. Her practice demands that she accept being caught between two sides.

These two traits have become constraints for the art of diplomacy. On the one hand, even if the relations of force are unequal, the exercise of diplomacy effectively requires that the question of a possible peace be posed as implying the demands of two "forces," even in an artificial manner. Without this, one ends up with the obscene confusion between a peace to be invented and a more or less honorable surrender. It is thus necessary that each party have the force proper to the idiot, who knows that there is something more important than a successful articulation. On the other hand, the diplomat is not empowered by the group to which she belongs, in the sense that she could make peace "in the group's name"; she is not its authorized spokesperson or plenipotentiary. She can bring back only a peace proposal that will be accepted or refused. That is why no diplomat is tempted to say, "In your position," to another diplomat. She knows that, like her, this other diplomat risks the accusation of betrayal. And the two of them know that they cannot share the risk that unites them. This risk, forced by the fact that, against all logic, "peace can be given a chance," gives the space-time of their encounter the character of the ancient Greek *kairos*, a putting in suspense of the probable. Diplomats are bound by the suspension that the groups who mandated them have consented to. Nothing more: with neither promise nor guarantee. But also nothing less: the *kairos* is the cause that renders them capable, despite being frightened of betrayal, of envisaging the possibility of an articulation where, logically, confrontation is demanded. The diplomat is thus always exposed to a double betrayal: the betrayal that she might be accused of and the betrayal of the *kairos* that obligates her to risk this accusation with other diplomats.

The parallels that I have made between the risks of the practice of the social sciences, such as I have envisaged them, and the risks of diplomatic practices, don't signify that practitioners of the social sciences would one day have to play a privileged role when it is a question of a possible peace. They could nevertheless have an important role to play, in the sense that their stories and experiences can contribute to the creation of a culture of diplomacy and its obligations. A culture that is all the more necessary, because when a situation emerges in which a war could perhaps be avoided, "diplomats" belonging to some all-terrain diplomatic corps should not intervene. Only those who belong to a practice, that is to say, those whom the obligations of this practice make hesitate, can be exposed to the accusation of betrayal, the risk of which is the signature of diplomatic intervention and of the peace proposal that arises from it. The capacity to produce diplomats when they are needed necessitates a culture of the imagination that the social sciences, as I envisage them, can nourish.

Far from being guaranteed, the possibility of a diplomatic success is therefore doubly conditioned. Not only is it necessary that, deciding against all logic to give peace a chance, each party be capable of mandating diplomats; each party must also be capable of hearing those it has mandated on their "return." In other words, each party must be capable of envisaging their diplomats' propositions and of coming to a decision about them.

One can understand why I have spoken of peace as "possible." "Diplomatic peace" has nothing to do with "the" solution, the dissolving in a wondrous ecological peace of the threat in the name of which the necessity of consensual pacification is invoked. The idea that pacification is the only stable, generalizable solution wasn't a misleading appearance. The frog of civil war is not going to become a charming princess, thanks to the kiss of ecology. What is dramatized is, on the contrary, precisely the absence of guarantee, that is to say, the gap between every general condition and every particular success. The practical culture associated with the ecology of practices is a general condition. No diplomacy is possible if a conflict is reduced to "they do not understand" or "they are blinded by their convictions" or "they are resisting change." But this "cultural milieu" doesn't guarantee anything; it only feeds the capacity to mandate diplomats and to listen to them when they return.

In both cases, what is decisive is the difference between identitarian appropriation and belonging. In effect, the possibility of a diplomatic success stems from its never being an obligation that foredooms war but always the signification conferred on an obligation, linking it to a requirement. For a diplomatic peace to have a chance of existing, it is therefore necessary that no discursive signification be held capable of defining what might be called the "meaning" (*sens*) of an obligation. This condition will receive the tranquil approval of linguists and philosophers, but something is added to it here that neither linguists nor philosophers usually anticipate: the distinction between meaning (*sens*) and signification cannot, must not, above all, lead to the general idea of the relativity of all signification. The questions that issue from the work of diplomats ("Is it acceptable?" "Does the signification it proposes for what obligates us respect the meaning of our obligations?") cannot be asked by the diplomats bringing back a proposal without fear and stammering. Certainly it requires that this proposal be heard, not rejected outright by an "identitarian reaction," a maintaining "on principle" of requirements that lead to war. But what is required is not a relativistic indifference, nor the application of a general rule that would permit the identification of significations which would respect or betray the meaning they claim to express. What is required is knowing how to hesitate as to the response to give, a knowledge that, as such, implies that practitioners are capable of hesitating with regard to what obligates them.

The manner in which such responses are produced, whatever the procedure might be, could be called "consultation." The advantage of this term is that it is neutral regarding this manner, as one can talk as much of consulting the people as of consulting spirits, and in all cases it implies a certain solemnity: it is not a question of "considering" one opinion against another but of convoking those to whom the response belongs.[11] Whoever the beings consulted are, what matters is the putting in suspense—everything that human language could accomplish has been accomplished, so something else, something that doesn't speak the same language, must now intervene among them.

One can propose here that the resources enabling a consultation are also those that protect the success proper to a practice from any identitarian appropriation. Knowing how to consult is also to know how to maintain the difference between obligations, which cannot be appropriated by anyone, which do not authorize any "and therefore," any certainty of deduction, and requirements, which do formulate such "and therefores." These resources would be pointless, and peace would be not only possible but guaranteed, if a practice didn't require anything. But then we would have shifted from an ecological problematic to the world of angels, those beings who, without bodies, neither affect nor are affected, but sing the praises of what they owe their existence to.

That practices have requirements and as a consequence impose the question of war is not a sad conclusion, the sighing of the disappointed utopia that is so easily transformed into denunciation or into a somber diagnosis of the voluntary servitude in which humans wallow. But the ecology of practices makes it matter that requirements do not have the last word. This ecology dramatizes as such the difference between the idiot and the sleepwalker. The idiot is someone who, knowing that there is "something more important," also knows that it is not up to him to take up the stance of the spokesperson of that cause-thing, even though it is up to him to hesitate: Is this a reformulation or a betrayal? This immanent test may be a response to the question of "sects and identitarian fanaticisms" because it selects those who know that what makes them think and feel doesn't belong to them. But such a test evidently does belong to the same (speculative) register as the notion of practice itself, and it makes explicit its correlate, the possibility of a practical culture of hesitation and consultation, of an affirmation by practitioners themselves, who do not know a priori what their belonging to a practice renders them capable of.

From this point of view, there is doubtless much to learn from the cultures that we have judged to be superstitious or fetishistic, these cultures that cultivate prudence, care, the art of paying attention, where we often demand binary differences, the good or the bad. They know that one must learn the manner in which beings are to be convoked or invoked and the appropriate manner to protect oneself when dealing with them. They know that what they honor is neither good nor bad but may be fearsome, must be given a *domus* and be nourished, so that it does not devour you. Modern practitioners judge such apprenticeship to be pointless, and they will have nothing do with such superstitions. But isn't that what then delivers them up, without protection, to everything that is then deciphered in psychosocial terms and is accounted for in terms of "human weakness"? Doesn't the sleepwalking invoked by the researcher working "under a paradigm" whom Thomas Kuhn put on stage or the horrified reaction "but that would open the door to . . ." of so many practitioners more generally testify to the power of what possesses them? Of what transforms them into predators, missionaries, or conquerors?

We started with the question: How is one to put "on the same plane" modern practices and those that would, for their part, live only on credulity and superstition? The answer is that it demands a great deal of work, not because of the credulity of the public but because such a plane demands an art of paying attention, a knowing how to hesitate, which is precisely what modern practices have unlearned. That is why this plane is speculative and why the ecology of practices bets on a possibility far from any probability. The only efficacy that it can hope for, as such, is the capacity to turn one's back on epic or apocalyptic rhetorics and to affirm that through the requirements that these rhetorics are the bearers of (and therefore!), what is at stake is not the "adventure of humanity" but the stupid and blind triumphalism of those who think they can be the incarnation of such an adventure.

THE COSMOPOLITICAL TEST

In 1996, when I was working on what was to become the *Cosmopolitics* series, the word *cosmopolitics* imposed itself out of the blue.[1] I knew that what urged itself on me in this way was the necessity of resisting the ancient Greek cosmopolitanism that wanted to be at home everywhere. However, I hadn't remembered that Kant had associated the same word with the project of a "perpetual peace." For Kant, the history of humanity was that of a progress that was both moral and juridical, making it conceivable that the barriers between sovereign states and between individuals were giving way to unification—when all would think of themselves as members of a world civil society, with the same rights and civil duties. When I realized this, I added some caveats to the manuscript. What I was not aware of, however, was that after the fall of the Berlin Wall, this Kantian use of the term *cosmopolitan* was in the process of reappearing, reinventing the utopia of a general progress of humanity that would find its expression in the authority of a *jus cosmopoliticum*. It's not impossible that had I known this, I would not have added those caveats to the manuscript but abandoned my use of the word.

No one is a proprietor of words, and one cannot require of philosophers that they drop them, abandon them, because of the possibility of a misunderstanding. But that the word *cosmopolitics* has reappeared, charged with the utopian drama of the destiny of "humankind," has made me decide to turn to it only at the end of this trajectory and to associate it entirely explicitly with what worried me when it first came to me. Because "cosmopolitics" was not synonymous with "ecology of practices" but responded to the necessity that I suddenly felt to slow down, faced with the possibility that in all goodwill I might be in danger of reproducing what I have been trying to think against with this ecology of practices.

For me, the term *cosmopolitics* will always remain associated with a cry that, so Alfred North Whitehead wrote, "echoes down the ages, *My Brethren, by the bowels of Christ I beseech you, bethink that you may be mistaken!*"[2] Whitehead called it "Cromwell's cry," and I discovered later that Cromwell was not addressing his Puritan brothers inhabited by an assured and vengeful truth; rather, he was threatening enemies whom he would later crush. Echoes are forgetful of their origins, and the efficacy of the cry has remained intact. This cry arouses fright in the mode of an indeterminate test. To be sure, Cromwell invokes Christ, but Christ here is not the bringer of a particular message: he is made present, but it is a presence without interaction, which doesn't appeal to any transaction or negotiation as to the manner in which it must be taken into account. Thinking in its presence calls for hesitation. I later learned that a shorter version of this cry serves as a quasi-ritual motto when Quakers meet in order to discern the meaning to be given to a concern expressed by a member of the group.

I was indeed "concerned" by the possibility that the ecology of practices might appear to be the "right solution," the solution capable of producing some sort of agreement. A concern of this sort might be said to be a little megalomaniac, as it presupposes that the propositions of a philosopher be able to produce agreement. But learning to think in a dangerous world—not in a world that would, through chaos and pain, be on track to the "right solution"—is what obligates me as a philosopher. It is thus not a matter of megalomania but of learning. As Leibniz would say, I thought I had reached port, and I found myself cast back out to sea.

We live in a dangerous world. One may think here of the old analysis of Joseph Needham, asking himself why technological inventions that China had absorbed without any ruptures could be placed at the origins of what was called the "industrial revolution" in Europe.[3] Many have said, and I still hear it being said only recently, that it was physics that made the difference, the great discovery of the fecundity of mathematics for describing the world. As an embryologist, Needham didn't stop at this point—he knew just how limited this fecundity was. For Needham, Galileo's work or that of Newton doesn't explain anything: it's the fact that each could be celebrated as an event that had to be explained. All the more so since each was considered quite coolly by Chinese astronomers. The explanation that Needham selected was one that emphasized the freedom that European "entrepreneurs" then benefited from, the entrepreneurs whose inheritors Bruno Latour has described, busying themselves with the construction of networks that get longer and longer, in defiance of all ontological stability, fearlessly tying human interests to more and more numerous and disparate nonhumans. Galileo was effectively the constructor of a network. He was able to turn into an event a knowledge that, when all is said and done, only concerned the manner in which smoothly rounded balls roll down a well-smoothed incline, to which can be added some observations through the telescope that allowed him to add some supporting arguments to the heliocentrism hypothesis. Galileo was able to create an event because he knew how to put this knowledge into direct communication with the grand question of authority, of the rights of enterprising knowledge in the face of philosophical and theological traditions. His condemnation was to no avail in a Europe that was fragmented between rival states and between rival religions, whereas in the unified empire that was China, he would doubtless not have been condemned; rather, widespread indifference would have stifled his entrepreneurial passion.

Networking involves what today are called *stakeholders*, those who have an interest in an enterprise and whom this enterprise connects, without unifying their interests. Today, there is nothing to stop these stakeholders, and a new political project, associated with the term "good governance," sets out to entrust them with responsibilities that representative democracy, defined as having "run out of steam," would no longer be able to shoulder. This is why, before coming onto the cosmopolitical question, I have first to slow things down and explicitly emphasize the dangerous possibility that what I have been calling "practice" might be assimilated into the "entrepreneurial position" that defines each of the *stakeholders*, whose association would be best fitted to managing an enterprise.

In the context of "good governance" what is to be understood by enterprise is everything that results from the association between interests judged legitimate.[4] We have been "prepared" for the spreading of this definition by the order-word according to which a private enterprise is always better managed than a public enterprise. A desirable world would emerge simply from the "mechanics" of the impediments that *stakeholders* can constitute for each other, and would do so for every enterprise, including those that were supposed to represent public interests. In order for what emerges to be optimal, it would be necessary, in addition, that the interest that each stakeholder defends be well-defined. To be sure, an interest can be open to everything that permits its advance (see the recruitment mechanisms described by Bruno Latour in *Science in Action*). But what is essential is that the *stakeholder* be an "opportunist," blind and deaf to the question of the world that his efforts contribute to the construction of. Everything is prey for the opportunist, and any question bearing on another mode of composing the world, any shift away from deafness to "Cromwell's cry," any concern for what is happening to this world can then be put in the same bag, because their effect is to diminish what it's a matter of maximizing.

In this dangerous world, it is quite probable that the "ecology of practices" could be confused by some with a (pointlessly) sophisticated version of "good governance." Entrepreneurial opportunism will stop at nothing. This was something that I learned when I discovered, with a quasi-surrealist feeling of horror, the manner in which *empowerment*, a theme that I had associated with the practices of American activists (in particular those seeking to make themselves capable of nonviolent action[5]) had been taken up a bit everywhere. What mattered now was the *empowerment* of *stakeholders*. Those with interests in a situation were to be freed from the constraining weight of rules that would prevent them from determining what is best for themselves. "We have the right to profit from our situation; we demand to be accorded the possibility of turning it to our greatest advantage"—this is what *empowerment* has become. No proposition can escape this fate if it can be hijacked to support the cause of good governance.

One manner in which one can protect oneself from any "recuperation" is, of course, to oppose to the "mechanical" composition of "blind forces" the antithetical model associated with the biology of the organism: democratic life can be assimilated to the harmonious participation of everyone in a single body.... The ideal of a harmonious composition can, in effect, be characterized as the "other" of the spirit of enterprise. But it's a dream that risks becoming a nightmare if it seeks to be realized, because it is limited to inverting the poles of the mechanical model. What does not vary is that the composition has no need of political thought, doubt, or imagination as to consequences. The body "knows best"; it is the cosmos, a realized cosmos; and woe to those in doubt. The capacity to hesitate, then, is no longer a force but a perversion, while intuition, instinct, the immediacy of sensing will be celebrated against the artifices of thinking.

It is not in the least bit certain that a living body functions in this harmonious mode, regulated by the coincidence for every part between "realizing itself" and "contributing to the common good." As long as biology makes the understanding of the body, or the organism, prevail over the entanglements of ecology, it will be dangerous for politics, because what has to be understood about the body, what its survival depends upon, is a relative stratification, what can be called a decoupling of scales (cells, organs, organisms). Thinking, imagination, hesitation then become the enemy, because they scramble the scales.

Neither physics, as a science of laws that verifies the command, "Obey nature in order to master it," nor biology, as a science of the modes of holding together on which the life or death of the body depends, helps in thinking the ecology of practices in its link with the question of politics. By contrast, the art of eighteenth-century chemists, who distinguished chemistry from mechanical calculation,[6] gives artifice, the creation of connections, the irreducible meaning that this ecology needs. Even today, chemists can hardly refrain from considering that the bodies they deal with are "active," gifted with an activity that cannot be attributed to them as an invariant property but rather as something that depends on circumstances. It behooves the chemist to create the type of circumstance that will enable these bodies to produce what the chemist desires. The chemical art—of catalysis, activation, composition, moderation—honors something we have a tendency to scorn: the manipulation that bends things in such a way that they "spontaneously" accomplish what the chemist wants to "make them do."

A strange model for politics, it will be said. But this sentiment of strangeness translates our idea that "good" politics ought to incarnate a form of universal emancipation: remove the alienation that separates humans from their freedom, and you will obtain something that resembles democracy. The idea of a political art, a "technique" even, then becomes anathema, an artifice that separates humanity from its truth. To refer to the art of the chemist is to affirm that there is nothing spontaneous about political association. This corresponds, moreover, to the question that John Dewey placed at the center of his work: How is one to "favor," to "cultivate," democratic habits?

If I am passionately interested in the notion of *empowerment* in the American activists' sense, it's because in answering Dewey's question, these activists have produced a whole culture of artifice that American activist witches have extended and called "magic." What has developed is an experimental creation of rituals that are efficacious in the production of *empowerment*, in the activation, thanks to the collective, of something that each member separately would have been incapable of producing. It's a radical art of immanence, because the efficacy that is sought is not at all submissiveness to the group or its objectives: on the contrary, it is the capacity to confer on a situation the group is confronted with the power to obligate this group to think together, that is, the capacity not to subject this situation to ready-made conflicting generalities. Where one might be tempted to speak of "belief" or "superstition," one can thus see an art of "activation," an art that, like that of chemists (and cooks), needs the learning of recipes, manners of proceeding that no general proposition has the power to anticipate.

But there is another aspect to the art of the chemist that can guide us: it is an art of heterogeneity, of the making present of bodies in their heterogeneity. This aspect is, in any case, also taken into account in the case of rituals the efficacy of which is to induce *empowerment*. These rituals put in place "roles"—like that of the serpent, who intervenes to complicate a discussion with a reminder about objections that have not been addressed—that form safeguards against the types of "spontaneous" understandings and misunderstandings that dominate meetings carried out with "goodwill." The emergence of a thinking of and in the situation requires an active resistance to the dynamics that are restricted to tolerating the always heterogeneous reasons for hesitating in order to make the "necessary unity" of the group prevail. Here also it's an art of artifice. There is an efficacy proper to the role itself, something that actors know well: the role is not just taken; it "takes" whoever assumes it. Even pilgrims ought not to feel insulted by the proposition according to which they have to assume a role and not simply allow themselves to be guided by faith.

The ecology of practices appeals to an art of artifice, to an art of the role that makes one think, feel, and hesitate, because it speculates on the difference between the sleepwalker—defined by a predatory, that is, an entrepreneurial, position—and the idiot, for whom what is "more important" than the general interest is not in the least bit a particular interest. The obligations that take practitioners, that force them to think, take them in the manner a role does. What fabricates a sleepwalker requiring the satisfaction of his interests is *identification* with a role—that is, its nonculture. The art of activating differences between requirements and obligations, which is what the ecology of practices demands, is pretty much exactly what would be condemned by the good governance of *stakeholders*, who gather together to dismember a situation in terms of their preexisting interests.

However, it isn't enough to think the ecology of practices against those who might see in it a means of short-circuiting the conflicts, complications, and constraints of politics. *We must also be wary of ourselves*, always so full of goodwill, so enterprising, always ready to speak for everyone else. This is where the term *cosmopolitics* arises for me, as an operator for slowing down.

Cosmopolitics is not synonymous with the ecology of practices but is born from the fright provoked by the manner in which a proposition like that of the ecology of practices could be accommodated in our unhealthy world. The danger is that what might be forgotten is the fact such a proposition is "signed": the history that it stems from, our history, is the same history that invented the category "politics." What it is a matter of slowing down, of making stutter, is the goodwill thinking according to which *at last* we would have a proposition "acceptable to everyone," a proposition that would convoke us all, not as "humans," for sure, but as "practitioners." Fright over the sudden proximity of this ecology with a new imperative: "Express yourself; state your objections; mandate your diplomats; make your manner of diverging exist; make your contribution to the common world that we are constructing matter." With a new [command]: practices and *empowerment* at every stage!

The cosmos, such as it figures in the term *cosmopolitics*, designates an inappropriable unknown. Coupled with the word *politics*, it has the type of presence evoked by Cromwell but addressed this time to those who are the masters of the "and therefore. . . ." Those who, in all goodwill, are subject to the temptation to become the representatives of problems about which they will say: "Whether we like it or not, these problems are imposed on us all and demand the participation of everyone." One could say that the ecology of practices is an operator of equalization, but as such it has a meaning only for those who are concerned by such equality. To be sure, equality here no longer requires a prior convergence, that of citizens defined by a common cause, but it mobilizes practitioners affirming their divergence. Nevertheless it is a matter of slowing down the "and therefore" that rushes in, once a solution starts to appear. The presence of the cosmos seeks to provoke hesitation, even fright, in the face of the possibility of this goodwill cropping up once again, crushing the murmuring of an "idiot" for whom there is *something more important than this mobilization*.

The murmuring of this particular "idiot" could be assimilated to the response of Bartleby the scrivener—"I would prefer not to"—in Herman Melville's celebrated short story. On the condition, though, that one focus not on Bartleby himself but on the narrator, the lawyer who happens to take on Bartleby as a copyist. Like the narrator himself, many commentators are fascinated by the figure of Bartleby, but it's the narrator who interests me, as it's his trajectory that is really frightening.

The figure of Bartleby carries out a passage to the limit: we will never know the meaning of the indifference that leads him finally to his death (he is imprisoned for vagrancy; he prefers not to eat). We can, however, understand the trajectory of the lawyer faced with this enigma. He collides with it. He is troubled, deeply troubled. He is ready to try anything, because he can't help feeling responsible. Yet he cannot let go of the rules of the social game that Bartleby disarticulates, either. Bartleby must agree to take on a role, whatever it might be. It will cost the narrator a fortune in diplomacy to propose a role that would be acceptable to Bartleby, and he will not even receive a refusal in return. Bartleby doesn't even contest the basic rules of the game. He offers no purchase, even for conflict. The desperation of the "chemist" for whom no proposition manages to "activate" something that obstinately, passively, remains a "foreign body."

But the crucial, properly frightening, point is that although the narrator seems to come to accept his failure, clients of the bureau take offense at the refusal of this inoperable scrivener, who prefers not to provide them with the services they demand. Called to account for his tolerance, the narrator will choose to vacate the bureau, as Bartleby preferred not to live anywhere else. But the narrator knows that the new occupants will do what he himself felt incapable of doing: force Bartleby to do what he prefers not to.

The lawyer's error, when others took offense, was doubtless not to have become the advocate of, not Bartleby himself (for how can one defend a "cause" one knows nothing about?), but rather the possibility that Bartleby, whatever his reasons might have been (and that is no one's concern), made him catch sight of: the possibility of letting be one who doesn't want to occupy any place, one who doesn't want to take any role. And that is doubtless what condemned him to vileness, to the decision to move offices, so as to be able to wash his hands of the fate of this unbending recalcitrant, knowing that others would resolve the question in his place.

The danger that the ecology of practices is the bearer of is that it finds itself—as the lawyer faced with Bartleby—faced with those whom one would like at all costs to make into interlocutors, producers (if possible) of constructive counterpropositions, at least of objections, whereas they prefer not to participate. It's a matter of resisting the urge to discover how to get them to take on a role, to express themselves, to manifest their divergence. To become an emerging public, in Dewey's sense. In fact, it is not a matter of addressing them, but of inventing the manner in which what our signature is—"politics"—might make its "cosmic double" exist, might be constructed "in the presence of" those who don't allow themselves to be enrolled.

CHAPTER NINE

There is nothing angelic about the ecology of practices; it's not a fairy tale in which all would find what is due to them in the construction of a "common world" that is good, at last. There are voiceless victims, just as there are species that silently disappear. From the cosmopolitical point of view, the offense is not that there are victims but that those who are doing the constructing don't accept the test that there being victims imposes. That— protected by their defense that victims "only had to participate"—they are not fully exposed to the price others may pay for their construction. The question here also is practical: How is one to be exposed, to be forced to hesitate, by what cannot, or does not want, or prefers not, to take part in the *calculemus* that is underway?

Perhaps victims need an advocate, something that the man of law was unable to be for Bartleby. But they certainly need "witnesses" able to make their presence exist, to try to channel what they cannot, don't want, prefer not to vindicate. Perhaps that's a role specifically suitable for those who are usually called "artists," because it's a matter of channeling something that is not of the order of a position, as a position has not been taken, but that belongs primarily to "sensation." Like "Cromwell's cry" echoing down the ages, the efficacy of sensation is to deprive our reasons of the power to disqualify what remains outside the terms of our calculation, to open it to the unknown that I associated with the prefix "cosmo."

Evidently, the presence of victims no more guarantees anything than does the staging of diplomacy: the cosmopolitical proposal has nothing to do with the miracle of decisions that "bring everyone into agreement." What matters is that forgetting, or, worse, humiliation, is prohibited. I'm thinking here notably of the contemptible idea that financial compensation ought to be enough to repair the undesirable consequences of a (good) decision, or the obscene attempt to divide victims, to isolate those who are recalcitrant by addressing mainly those who, for one reason or another, will come to an agreement more easily. Those who attempt to produce an agreement ought to know that nothing will be able to erase the debt that links their possible decision with its victims.

We are a long way now from the art of chemists, because the presence of victims has nothing to do with what chemists call a "buffer" solution, which keeps a reactive milieu in a zone defined as optimal, with neither too much acidity nor too much basicity. However, thinking and making decisions in the presence of victims does correspond to a protection against what threatens us, the temptation to seek a guarantee as to the difference between what must be taken into account and what one has the *right* to neglect. It is this role of guarantee that "science" has been asked to play, in the name of "reality," but this same role could be played by politics, when only those who "feel concerned" by an issue have the legitimacy to decide how it will matter.

That is why I will finish this essay with the words of neopagan witches, who have searched for their strength within the active memory of an eradication that nothing will ever "compensate." Words that channel destruction, rape, the brutality that dismembers, that make it felt in order to metamorphosize them into a power of feeling and thinking.

> Breathe deep
> Feel the pain
>> where it lives deep in us
>> for we live, still,
>> in the raw wounds
>> and pain is salt in us, burning
> Flush it out
> Let the pain become a sound
>> a living river on the breath
> Raise your voice
> Cry out. Scream. Wail.
> Keen and mourn
>> *for the dismembering of the world.*[7]

THE FIRST EXPERIMENTAL APPARATUS?

What follows is a personal hypothesis regarding what was perhaps the first experimental demonstration, the demonstration that, in 1608, turned Galileo into a "modern scientist." It is a hypothesis that is compatible with the data, including the fact that the first steps in the demonstration haven't left any written trace. It was made concrete by an installation offered to visitors of the exhibition *Laboratorium*, which took place in Anvers between June 27 and October 23, 1999. Didier Demorcy conceived and brought about the installation.

Installation Description

In a large room, three tables are set around a circular box of sand. On each table, two inclined planes, neither fixed, have been placed. Steel balls, a measuring grid, and two metallic plates, each of about one square meter, complete the equipment for each table. Each inclined plane permits a variable incline. The written text that follows was available for visitors to take away with them.

From the Motion of the Balls to the Motion of the Earth

I. HOW THE MOVEMENT OF FALLING BODIES BECAME A STAKE · What if the Earth rotated on its own axis and around the Sun?

Published in 1543, Copernicus's hypothesis provoked a cultural and theological scandal. It went against scripture: wasn't it the Sun, and not the Earth, that Joshua ordered to stop moving so as to give him the time to win a decisive battle?

But the motion of the Earth comes up against a more formidable obstacle. How can the Earth move without our being aware of it? Here it's a physical motion, that of falling bodies, that lies at the center of the debate. Clearly, heavy bodies, like apples falling from a tree, meet the ground by the shortest route possible, that is to say, by a straight line. But if the Earth—that is to say, the ground—moves while an apple falls from a tree, this apple, falling in a vertical straight line, ought not to hit the ground at the foot of the tree. It ought to hit the ground where the tree would have been if the Earth had stood still. Thus, if the Earth rotates, we wouldn't see bodies fall in straight lines, but obliquely!

The motion of falling bodies thus became a stake. This motion seems to testify that the Earth really is immobile.

2. THE 1608 SCHEMA · In 1608, Galileo, at the age of forty-four, was teaching at the University of Padua. He was already highly respected as an astronomer, but he was best known for his work on motion and the equilibrium of bodies. For some years he had been pursuing a single problem: How to describe the manner in which bodies acquire or lose velocity?

To define the velocity of a uniform motion is easy: it is the relationship between distance traveled and the time it takes to travel that distance. But when the motion accelerates, velocity is changing all the time. And how is one to define the velocity of a body if it doesn't travel a measurable distance at a given velocity? Galileo tried for years to answer this question, but without any success.

In 1608 he sketches out a schema and lines up the numbers. The numbers allow the schema, and the situation that it represents, to be understood. Galileo lets balls roll down an inclined plane placed on a table. For each height of the starting point, he measures the distance between the point of impact of the ball on the ground and the edge of the table.

The page from 1608 marks an event.

It is, effectively, the trace of an experiment that Galileo would not have been able to invent if he hadn't already imagined the answer he was seeking. He had already worked out what it was that, according to this hypothesis, he "would have to" measure (a series of numbers mentions *doveria*, "would have to"). In other words, the experiment is a sort of rendezvous proposed to the motion of the falling body by Galileo: if the numbers are what they "ought" to be, the hypothesis will be confirmed.

But the inclined plane doesn't just allow Galileo to confirm his hypothesis. It gives him the power to prove it. The inclined plane gives motion the power to impose—against all objections—the manner in which it must be described. Thanks to the inclined plane, Galileo in effect doesn't simply describe motion: he stages it in a mode that allows him to vary, separately, the different factors that can intervene in it. In other words, the inclined plane allows motion to be understood as a mathematical function—that is, in terms of variables.

It can be said that the page from 1608 is a trace of the first experimental demonstration. And therefore that the inclined plane is the first experimental apparatus, transforming a phenomenon into an instrument of proof.

Evidently, in 1608 there was nothing new about making balls roll down a slope. It was Galileo's gestures, gestures that you yourselves can make once again, that inaugurated [this] new history.

By trying out Galileo's gestures, you will encounter what Galileo himself perhaps experienced in 1608: the force of the laboratory . . .

3. THE FIRST LABORATORY? · The great richness of the laboratories that followed after Galileo lies in the ensemble of apparatuses that, like the inclined plane, have succeeded in transforming what they interrogated into reliable witnesses, imposing the manner in which they must be described. Each one of these apparatuses has thus been able to become a measuring tool that makes possible the invention of a new apparatus. But for the first laboratory to be able to demonstrate, it could not, for its part, presuppose anything: it had to be transparent. Galileo's inclined plane did not presuppose any other apparatus, any other instrument or measurement, except the measurement of distance, which had been around for millennia. In particular, Galileo succeeded in interrogating motion without having to measure time.

Go into a laboratory today and ask about the functioning of all the instruments that those who work there need: little by little, it is the entirety of the experimental science of the past that will have to be deployed. In doing this, you will be going back in time . . . all the way to 1608.

4.1. WHAT THE POINT OF IMPACT TELLS US · The distance between the point of impact and the edge of the table varies as a function of the ball, from the moment that it leaves the table, and this depends solely on the velocity that descending the length of the inclined plane gives it.

If you doubt that this is the case:

+ Release a ball at different points on the inclined plane. This allows you to vary the velocity of the motion of the ball rolling along the table. And you will note that the faster this is, the greater the distance between the point of impact and the edge of the table.
+ Change the position of the inclined plane so as to vary its distance in relation to the edge of the table. For the same starting position of the ball on the inclined plane, the point of impact stays the same, regardless of the distance the ball has to travel across the table. Motion on the table thus has no measurable consequences: it doesn't make the velocity of the ball change.

Be careful: for this to work, the ball has to be released, not thrown. If at the outset, you give it some impetus—that is, a velocity—nothing will work . . .

4.2. WHAT THE MOTION OF THE BALL ON THE INCLINED PLANE TELLS US · The velocity that the ball gains depends on the height of its starting point on the inclined plane alone. It therefore does not depend on the velocity of its descent down the length of the inclined plane—that is, on the time it takes to complete the descent.

If you doubt that this is the case:

+ Vary the degree of incline, the steepness, of the inclined slope, but always release the ball at the same vertical height (measured from the horizontal plane formed by the table). The time the ball takes to descend and the distance it travels down the inclined plane can vary as much as you like, but the point of impact will not change.
+ Vary everything as much as you like. If you want two motions to produce a point of impact with the same distance, there is one, and only one, solution: to start from the same height.

The velocity that the ball gains doesn't depend on the length of the slope it travels down the inclined plane, nor on the angle of its incline, nor on the time the descent takes, but solely on the variation of height! When it is a question of velocity gained (or lost), "time doesn't count." It will take more than a century for the ensemble of learned Europeans to get used to this "experimental fact" and understand its importance.

The first part of this experiment made use of the free motion of the ball, resulting in its impact with the ground, so as to identify what the velocity of the ball, at the moment it leaves the table, depends on. It made it possible to prove that a single variable determines the point of impact: the height of the starting point. This variable will now become a (reliable) instrument for the varying of velocity, allowing free-falling movement, from the table to the ground, to be characterized. The apparatus thus articulates two modes of measurement, according to whether free fall motion is doing the measuring or is the object of measurement.

4.3. WHAT THE MOTION OF THE BALL FALLING OFF THE TABLE TO THE GROUND TELLS US · The motion of the ball in free fall from the table to the ground can be understood as the sum of two motions: the ball follows its horizontal motion at uniform speed across the table, and, at the same time, it adopts the same, accelerated vertical motion it would have adopted if it was falling vertically. The two motions combine without influencing each other: the ball continues to move forward while falling.

If you doubt that this is the case, try this demonstration, which requires two people and occurs in two stages.

+ Release a ball at a determinate height on an inclined plane and measure the distance to impact. Position another inclined plane, with the same degree of incline, on the table such that the distance between it and the edge of the table is equal to the sum of the impact distance that you measured and the distance between the first inclined plane and the edge of the table. Position a metal plate on the ground so as to "detect impact by sound." Release two balls at the same time, one on each plane and at the same determinate height as before. At the moment that that first ball hits the ground, the second reaches the edge of the table (you can verify this more precisely by placing a second metal plate vertically at the edge of the table to block the movement of the second ball). Thus the two balls have traveled the same distance horizontally, in the same amount of time, one on the table, the other in falling.

+ Leave the metal plate on the ground. Release a ball down an inclined plane. Have a second ball at the ready at the edge of the table, ready to drop. When you see the first ball reach the edge of the table, release the second ball. If your gesture is precise, both balls will hit the ground at the same time: whatever the horizontal component of its speed might be, the ball takes the same time to reach the ground.

In both cases, one has to learn the knack of releasing the balls to allow for synchronicity. But it is the way that the simultaneity of impacts is not erratic, but improves as your gesture becomes more precise, which provides proof.

4.4. TO TAKE THE EXPERIMENT FURTHER · Measure the distances of impact with the ground for different starting heights, without changing the inclination of the plane. If everything goes well, you will find that these different heights are more or less the squares of these distances. That is what the column of numbers states in Galileo's schema, and it's doubtless that relation that Galileo wanted to establish. This experimental relationship supposes and verifies a hypothetical mathematical definition of accelerated falling motion that Galileo had formulated in 1607: "in equivalent times they receive equivalent moments of velocity."[1]

The relationship established by Galileo, thanks to the inclined plane, can verify this hypothetical definition only if it is admitted that at any instant whatever [quelconque] of its descent, the velocity gained by the ball on the inclined plane depends solely on the height that it has descended at that instant. Galileo knows how to evaluate the velocity gained by the ball at the end of its descent, at the instant it moves from the inclined plane onto the table, but he can affirm that it is an indifferent [quelconque] instant in the descent, because the experimenter can choose it at will, by making the starting point of the ball vary.

5. WHAT GALILEO SUCCEEDED IN DOING · Before Galileo, free fall motion had a well-determined identity. The 1608 experimental apparatus "plunges" this motion into a continuous set of possible modes of descent, which differ as much from the point of view of the time taken to descend from a given height as from the point of view of the distance traveled.

With the inclined plane, the time taken by descent becomes a function of the incline of the plane. Vertical free fall corresponds simply to the minimum time (for an incline of 90°). The other extreme could be defined by an incline of 0°: the ball remains immobile on the horizontal plane. But one can henceforth conceptualize the possibility of a slope characterized by an extremely shallow incline, such that the ball would take centuries to descend a single centimeter. The plane would have to be extremely long, and it would have to be perfectly free of friction: thus it forms a "thought experiment," but one can, "in thought," determine the velocity that the ball would gain after descending for one hundred years, for example . . .

The motion of vertical free fall, with the ball landing at the foot of the table, becomes a particular case, which corresponds to a horizontal velocity of zero at the moment at which the fall begins. The inclined plane is then an apparatus that, depending on the starting height on the plane, allows a well-determined horizontal velocity to be conferred on the ball at the moment it starts its descent. The vertical becomes a particular limiting case of a family of parabola.

6. WHAT GALILEO SUCCEEDED IN SHOWING

+ Some people will perhaps remember learning the following formula at school: $mv^2/2 = mgh$, where m is the mass of the body in motion and g is gravitational acceleration. It signifies that, at every instant of its descent, the velocity, v, gained by a falling body depends solely on the height descended, h. In other words, one can say to Galileo's ball: "Give me the height value (h) of your descent, and I will tell you how much velocity (v) you have gained, whatever trajectory your descent followed and however long it took to follow this trajectory."

This possibility, of ignoring the details of the path taken in favor of a single piece of information, was a great surprise, and at the end of the seventeenth century, some people still hadn't accepted it. For us, it is what constitutes the extraordinary privilege of the (dynamic) motion on the basis of which physics formulated its laws. But one must be careful, because this privilege only holds up if there is no friction. The experimenter has to arrange matters so that any friction is negligible: the plane must be as smooth as possible and the ball perfectly round.

+ Among the beings that Galileo succeeded in defining, the most singular is the pendulum, the ancestor of techniques that finally allowed intervals of time to be measured precisely. Just as the velocity of the falling body depends solely on the height of the descent, so the periodicity of the pendulum swing (at a given point on the Earth's surface) depends solely on the length of the pendulum's cord. A first "scientific" measuring device is born.
+ Bodies do not fall "naturally" in a vertical line. If they are moving at the moment they are released, they will retain this movement with the same velocity and direction while they simultaneously fall vertically with a uniformly accelerated velocity. This is why, generally, projectiles follow a parabolic trajectory as they fall to the ground, combining a uniform and an accelerating motion. A body that falls according to a vertical

trajectory corresponds to the particular case in which the body is immobile when it starts to fall.

⋆ Above all else: the ball spoke! Before Galileo, the manner in which bodies fall was a stake of many interpretations and speculations. After Galileo, it is known that sometimes it is possible to give facts the power of imposing how they have to be described.

7. HENCEFORTH THE EARTH CAN BE IN MOTION WITHOUT US REALIZING ... · Galileo first heard of the telescope in 1609. In 1610 he published his observations of the Moon and the satellites of Jupiter in *Siderius Nuncius* (*The Starry Messenger*) and presented them as arguments: the Earth is just a planet, like Jupiter. The fiercest defender of Copernicus was born.

But it was perhaps the modest balls rolling down a humble inclined plane that authorized Galileo to think that he could demonstrate the truth of the motion of the Earth.

Because this hypothesis in astronomy came up against terrestrial evidence. If the Earth moved, how is one to explain, for example, that the apple that falls from the apple tree hits the ground at the foot of the tree? Now, though, Galileo could reply: even if the Earth was in motion, the apple would fall at the foot of the tree. Or, more precisely, its fall would be vertical from the point of view of the apple tree and from our own point of view, as we look at the apple and the apple tree.

In effect, Galileo's apparatus allowed him to show how the manner in which a body falls depends on its velocity at the moment that it starts to fall. Now, if the Earth is in motion, we all share its movement: not just the observer and the apple tree, both solidly planted on the ground, but also the apple attached to the branch. And when the apple detaches itself from the branch and falls, as it falls it will continue to displace itself with the same movement as the Earth and the apple tree. Exactly as the ball, at the moment that it leaves the table, continues the motion that it had on the table, although it is falling to the ground. The apple thus accompanies the motion of the apple tree and falls at its foot.

By contrast, an extraterrestrial observer, who could observe the motion of the Earth without sharing in it, would be able to see the apple following a superbly parabolic motion, similar to the parabolic movements followed by balls falling off the table after rolling down the inclined plane. Similar also to the movement followed by a bomb falling out of a plane in motion.

Galileo demonstrated that the Earth could be in motion: the falling of bodies was no longer opposed to this. But it was another astronomer, Johannes Kepler, who really put the Earth in motion.

Since antiquity, all astronomers, including Copernicus and Galileo, had been guided by one conviction: only the perfection of the circle allowed the perfect regularity of the skies to be understood. In 1609, Kepler dared to "break the circle." He showed that it was the ellipse and not the circle that allowed for the mathematical description of the movement of the planets and the observational data of astronomy to be brought into agreement. The Sun is not in the center of these movements but is situated at one pole of an ellipse.

Since Kepler, astronomy obeyed a new requirement of perfection: no longer the perfection of the circle but the perfection of the most precise possible agreement between mathematics and observation.

Galileo never recognized the importance of Kepler's discovery.

8. CONSTRUCTING THE PARABOLA · A ball leaves the table with a horizontal directional velocity of V. This signifies that it travels a horizontal distance, x, in time, t, with this velocity, per the formula $x = Vt$. Its vertical descending motion, for its part, responds (as a consequence of the fact that in "equivalent times they receive equivalent moments of velocity") to a formula that nowadays we write $y = gt^2/2$, where g is the gravitational constant and y is the vertical displacement effected after a descent time of t.

At each instant of the descent, the ball's position is thus given by two numbers: one that measures its horizontal displacement, and one that measures its vertical displacement. The parabola is the response to the question: For a horizontal displacement value x, what is the vertical displacement value y?

For any horizontal displacement value whatever, x, the ball must have been in movement for a time value of $t = x/V$. During this time, we know that it must have traveled a vertical distance of value $y = gt^2/2$.

$$y = gt^2/2 = g(x/V)^2/2 = gx^2/2V^2$$

As g and V are constants, $g/2V^2$ is a constant, and one arrives at the simplest particular case of the definition of a parabola, $y = ax^2$, which responds to the canonical formula $y = ax^2 + bx + c$.

QED

NOTES

Translator's Preface

1. "Vierge" without the article in French usually means "Virgo," as in the star sign, and the linguistic construing of intimate address is a bit trickier.

2. For a good example of this, see the final section of the chapter "Justifying Life?" in Stengers's *Thinking with Whitehead*, which offers a discussion of William James in lieu of a discussion of Gilles Deleuze and Félix Guattari's theory of desiring production.

3. As well as being a demanding author, she is a translator herself and has commented on her experience of this process in relationship to translating the works of Donna Haraway.

4. In her discussion of the etymology of the word *version*, Despret points out that in the sixteenth century, the word started to signify translation, or the necessity of "bringing something from another world, which is going to become part of our own." *Version*, in Despret's construal, concerns the becoming both of a text and a world. See Despret, *Our Emotional Makeup*, 22–23.

5. *Réactiver le sens commun: Lecture de Whitehead en temps de débâcle*, published in 2020, is a differently articulated version of a 2017 book *Civiliser la modernité? Whitehead et les ruminations du sens commun*.

6. I've borrowed this expression from Stengers's discussion (with Bruno Latour) of the work of Etienne Souriau. See Stengers and Latour, "The Sphinx of the Work."

7. One of the reasons *Capitalist Sorcery* is subtitled "Breaking the Spell" is to avoid an uncomfortable translation of *désenvoûtement* as "disenchantment." "De-enchantment" is the ugly, cumbersome, but more accurate, alternative.

8. This is not to say there is never provocation: see the discussion of the choice of expression *capitalist sorcery* in Pignarre and Stengers,

Capitalist Sorcery, 39–40. The target is very different from the indifferent "humanity and its illusions" that iconoclasts have often taken aim at. See also the discussions of "narcissistic wounds" in, for example, Stengers, *Hypnosis between Science and Magic*.

9. In the entry "Progress" of *100 mots pour commencer à penser les sciences*, written with Bernadette Bensaude-Vincent, Stengers writes of the "well-anchored habits that make the order-word 'progress' rhyme with the permission not to think."

10. Stengers, *Hypnosis between Science and Magic*.

11. "Our guardians" ("nos responsables") is a figure discussed in some detail in chapter 2 of Stengers, *In Catastrophic Times*.

12. Stengers, *Réactiver le sens commun*, 25.

13. An option is genuine, in James's sense, when it is "of the forced, living, momentous kind." James, "The Will to Believe," in *Writings of William James*, 718.

14. Stengers, *Réactiver le sens commun*, 191.

Chapter 1. Scientists in Trouble

1. Writing "men and women" is [not only] heavy-handed [but problematic today]. Faced with the question of "gender" now imposed on writers in the French language, I choose to follow the path invented by Anglo-Saxon women, that is to say, the arbitrary use of the feminine from time to time. The surprise effect seems more adequate to the sought-after goal than the heavy-handedness of the duplication "men and women."

2. Sokal, "Transgressing the Boundaries," 217–18.

3. The opening lines of *The Invention of Modern Science* (originally published in 1993 with La Découverte as *L'invention des sciences modernes*) predicted this conflict. I affirmed that "the thinkers of science sharpen their weapons [and rise to the defense of a threatened cause]" without knowing how right I was (Gross and Levitt were already at work), simply because it was inevitable. Stengers, *Invention of Modern Science*, 3.

4. I won't talk here about the relations between the Bush Administration and scientists concerning those questions to which the "fundamentalist" Christian electorate is sensitive, as well as those that annoy industrialists. It is a situation that is too simple, too much of a caricature to provoke anything other than the raised shields of a deceptive unanimity.

5. Dominique Pestre defines such a regime as an "assemblage of institutions and beliefs, practices and political and economic regulations, which delimits the place and mode of being of the sciences." The mode of valorization of scientific knowledges forms part of the assemblage. Pestre, *Science, argent et politique*, 36.

6. One could speak of this regime as "modern" in two senses. On the one hand, all the protagonists present themselves as modern, in the service of progress, as the traditional powers are no longer in a position to threaten anyone at all. On the other hand, the grand narrative of the advancement of disinterested knowledge is the sign of a modernity that proponents of "postmodernity" make it their joyful duty to tear to pieces.

7. On this subject, see, for example, part II, "Learning to Protect Oneself," in Pignarre and Stengers, *Capitalist Sorcery*.

8. Serres, "Vie, information, deuxième principe," in *Hermès III*, 72.

9. This attempt, which began with *The Invention of Modern Science*, is organized around the question of practices, with the notions of requirement and obligation by which that question is articulated, throughout the seven books of my *Cosmopolitics*.

10. That such a creation is not only possible but likely to nourish the links between researchers and their own practices is something I experienced in recent years, in the context of an interuniversity research project, "Les loyautés du savoir: Les positions et responsabilités des sciences et des scientifiques dans un état de droit démocratique" (a project financed by the Belgian Federal Science Policy Office). I am grateful to the research collective associated with this project, who allowed me to enrich, to better understand the scope of, and to put to the test the notions presented in this book.

11. Pierre Le Hir, "La société en mal de science," *Le Monde*, December 22, 2004.

12. The Virgin Mary will appear only in the final third of this text. The role that she will play there, in so far as her appearances designate places of pilgrimage, owes everything to Elisabeth Claverie, whose work on this topic has helped me to leave the territory defined by scientific practices. During a presentation, I have seen Claverie be submitted to questioning that implied that her way of approaching the pilgrims' experience meant that she, too, could be suspected of being "a believer in the Virgin Mary." By not creating an a priori difference between those who believe and those (like her) who know, nor by operating through "empathy," sharing the suffering and the hopes of those she studied, her way of approaching her topic did not

"insult" pilgrims. It is to the professional "putting herself at risk" in order to "describe well" the role conferred by pilgrims on the Virgin Mary that I pay homage. But this homage does not of course compromise Claverie in the theses that I am going to develop.

13. The "absent" or "crossed out," the unknowable of negative theology, is all that remains—for those attached to it—when this dismissal has cleared away everything else.

14. Stengers borrows Deleuze's expression "penser devant [les analphabètes, les alcooliques, etc.]" where "devant" has the sense of being in front of, in the presence of, before (as in "appearing before an audience"). "In the presence of" is a reasonable approximation as long as one bears in mind the complexities of what "presence" might mean here. [TN]

Chapter 2. The Force of Experimentation

1. In the text, Stengers refers to "la" science, to be distinguished from sciences and scientific practices. Where she does so I've taken the liberty of rendering this with a capital letter to indicate that we are in the domain of rhetoric referred to in the previous chapter. [TN]

2. On this subject, see the fine book by David Lapoujade, *William James: Empirisme et pragmatism*.

3. "Project Poltergeist," *Horizon*, BBC, broadcast on March 18, 2004, https://www.bbc.co.uk/science/horizon/2004/poltergeist.shtml.

4. The French *mot d'ordre*, which Stengers uses here in a manner analogous to that of Gilles Deleuze and Félix Guattari, has been translated throughout variously as "slogan," "command," or—as with Deleuze and Guattari—"order-word," depending on what it is a matter of accentuating. [TN]

5. For a eulogy of the version, see the introduction to Despret, *Our Emotional Makeup*.

6. See Latour, *Making of Law*.

7. The two registers to which the term *cause* refers intersect with the distinction between intermediary and mediator proposed by Latour, *We Have Never Been Modern*. The intermediary, like the relationship of cause and effect, gives primacy to the homogeneity of the terms that it links. It can be described in terms of a transmission, the ideal of which is fidelity. But a mediator is operative, and the operation is not defined in terms of faithful transmission, because the terms of

the bringing into relation that is brought about did not preexist it as such. No mediation is arbitrary; nor does any have a general value. Every mediation implies the definition of what, in this case, will be a success, and it is this definition that constitutes a "cause" obliging those whom it engages.

8. "Vouches for the responses," in French, is *répond des réponses*, and a "guarantor" is a *répondant*. [TN]

9. William Van Orman Quine, sacralized by American philosophers as their greatest, arrived at this judgment, which is essentially no different from those that triggered the "science wars," but he didn't provoke the anger of the (few) scientists who read him, because he affirmed that he respected science but was not able, as a philosopher, to understand it any other way. This enabled scientists to conclude that "philosophers understand nothing." Thus in *Dreams of a Final Theory*, 21–22, Steve Weinberg writes, "Ludwig Wittgenstein, denying even the possibility of explaining any fact on the basis of any other fact, warned that 'at the basis of the whole modern view of the world lies the illusion that the so-called laws of nature are the explanations of natural phenomena.' Such warnings leave me cold. To tell a physicist that the laws of nature are not explanations of natural phenomena is like telling a tiger stalking prey that all flesh is grass. The fact that we scientists do not know how to state in a way that philosophers would approve what it is that we are doing in searching for scientific explanations does not mean that we are not doing something worthwhile. We could use help from professional philosophers in understanding what it is that we are doing, but with or without their help we shall keep at it." Weinberg's book belongs to a key moment: Weinberg pleads the case for the supercollider, which he refuses to imagine could be canceled by the US Senate. Nevertheless, that is what the Senate decided to do, and Weinberg becomes one of the main protagonists in the science wars.

10. "Project Poltergeist,'" see note 3 above.

11. There's no dancing in biotechnology laboratories for a successful genetically modified organism. The "injection" of a fragment of DNA and an experimental success do not coincide. The "successful" GMO arises from a sorting operation in which the only organisms that are kept are those whose genetic modification entails the sought-after consequences. One can in this respect speak of a "small biology," a biology that is certainly sophisticated but fundamentally "instrumental," in the sense that the reason for the difference between the cases in which "it worked" and those where "it failed" doesn't matter much.

Chapter 3. Dissolving Amalgams

1. See Rose, "My Enemy's Enemy," 61–80, in the expanded version of the famous special issue of *Social Text*, with Sokal's "hoax" text removed. The business of the hoax gives Rose's title a premonitory quality. Having passed off as "sincere" a text that Sokal himself judged to be ridiculous, the hoax was possible only because a hazy text authored by Allan Sokal, a professional physicist, was judged worth publishing, doubtless because of the fact that the physicist was changing camps and had to be welcomed as a "friend," as he would provoke the anger of his colleagues and turn them into enemies.

2. The Radical Science Movement was born in 1975. A movement for the struggle of British scientists (a different movement existed in the United States) to promote the social and political responsibility of scientists, to denounce the harms committed in the name of science, such as IQ tests and genetics, and of course to promote the collaboration with other groups struggling against nuclear weapons, the arms race, and so on.

3. Collins and Pinch, *Golem*, 146.

4. Collins and Pinch, *Golem*, 2.

5. Collins and Pinch, *Golem*, 2.

6. Rose, "My Enemy's Enemy," 76.

7. Today a similar tension exists with regard to evolutionary psychology. The critical sociologists who have since proliferated are often in the camp of those who snigger with regard to the possibility of defining a "good science."

8. Rose, "My Enemy's Enemy," 70.

9. See Schatzki, Knorr Cetina, and von Savigny, *Practice Turn in Contemporary Theory*.

10. This is why Michael Polanyi, Ludwig Wittgenstein, Martin Heidegger, and Hubert Dreyfus, united by the fact that they have all emphasized the importance of tacit "know how" irreducible to propositional knowledge, are cited as "predecessors."

11. The very project of describing these practices as a naturalist may describe the behavior of ants—a behavior that can have its reasons but reasons that the naturalist doesn't have to share—is a classic in the human sciences. But here it is exacerbated by the fact that the scientists are interested in sharing their reasons. Can one imagine an informant from Africa addressing his ethnographer and saying to him, "Okay, you demonstrate an 'empathy' for our practices, but do

you really accept the foundedness of the interpretation according to which my father has been the victim of witchcraft?" We'll not go any further in this direction here: let it suffice to say how fortunate it is for ethnography that the category of "belief" belongs to the tradition that gives ethnographers their mandate, not to that of the peoples who host them.

12. In Schatzki, Knorr Cetina, and von Savigny, *Practice Turn in Contemporary Theory*, this position, inspired by Foucault, is defended by the philosopher Joseph Rouse, who equally underlines his proximity to feminist standpoint theorists; see Rouse, "Two Concepts of Practices," 198–208. Like them, he affirms the characteristic inseparability of a knowledge and the explicit, constructed commitment that gives it its relevance, in contrast to those who conjugate rationality and neutrality. I have borrowed my use here of the adjective *positivist* from Rouse, designating an address to the sciences that endows them with an identity that ensures for those who can explicate this identity the "sovereign" position of "being able to define one's object."

13. This may bring to mind what I would call the "first Latour," the Latour who, nearly thirty-five years ago, wrote *Science in Action*, whom Donna Haraway questioned for his "Machiavellian" approach. And the acronym for the Actor Network Theory associated with him, ANT, is not entirely contingent, even though, here, it is the sociologists themselves who are the ants, tirelessly following networks. This is the moment to recall that a certain number of other Latours were rapidly not substituted for but added to this first one, notably the Latour who, in a mode that cannot arise from positive sociology, dared to propose the "we" of *We Have Never Been Modern*.

14. See Stengers, "The Science Wars," in *Cosmopolitics*, 1:1–83.

15. Poincaré, *Value of Science*.

16. See the appendix to this book, where I situate the point at which Galilean "mathematization" of movement is anchored at the level of experimentation itself: Galileo's successful experimental staging was equally the confirmation of a "functionalization" of what he interrogated, making explicit its dependence with regard to what then have the status of variables, in a mathematical sense.

17. Perrin, *Atoms*, 206–7.

18. See the fine article by Carlo Ginzburg, "Clues: Roots of an Evidential Paradigm," in Ginzburg, *Clues, Myths, and the Historical Method*, 87–113.

19. Devereux, *From Anxiety to Method in the Behavioral Sciences*, 31.

20. See Chertok and Stengers, *A Critique of Psychoanalytic Reason*; and Stengers, *Hypnosis between Science and Magic*.

21. Thus, as Latour has shown, superbly well, in *The Making of Law*, what makes jurists hesitate is not at all that a case will allow them to learn something new, but on the contrary, what will allow them to "recognize" it, to link it to other cases.

22. See Pignarre, *Qu'est-ce qu'un médicament?*

23. Stengers is playing on the proximity between the verb *éprouver*, meaning to test, and the verb *prouver*, meaning to prove. An *épreuve* is a test and an ordeal, something that is trying. [TN]

24. It's because the industry, which has been entrusted with putting the clinical trials to work (the state restricting itself to receiving completed reports), becomes a bit "too" active—selecting, redefining indications and thus groups, even inciting the creation of new classes of pathology corresponding to a "subgroup" for which the verdict may seem to be positive, and so forth—that it can be accused of cheating: it doesn't submit to the convention. It infringes it.

25. The more and more glaring defects of clinical trials do not primarily concern the long-term effects of the medication or the effect of it being taken at the same time as another medication (patients in the statistical sample are not supposed to take another medication), however urgent those questions may be. They concern the nonrespect of the convention by industry—and the astonishing naïveté of the other protagonists (who perhaps believe in the "proof") that puts clinical trials into crisis.

26. See the intervention of patient associations into the conducting of clinical trials for AIDS medication described by Steven Epstein. Epstein, *Impure Science*, 295–328.

Chapter 4. *The Sciences in Their Milieus*

1. One might say that the manner in which Kepler felt obliged to attempt to make a faithful mathematical "portrait" of the solar system on the basis of observational data announces the singularity of a celestial mechanics. From the technique of perturbations to the laborious possibilities of "making" an unmanageable system of equations "speak" (at the outset, Urbain Le Leverrier's calculation, which enabled him to "find" Neptune, involved 269 equations with 13 unknowns!), from the three-body question to the question of chaos today, from Harvard's "computer" women to today's megacomputers,

this field of research has always worked "at the limits," putting to work every possibility of making the most faithful mathematical description of the awesome complexity of "this" planetary system. Celestial mechanics is, in this sense, closer to a "naturalist" science, seeking to describe an existing being in its insistent particularity, than to an experimental science, which is only interested in a situation to the extent that it is capable of verifying a hypothesis.

2. Thanks to the help of Didier Demorcy, I have had the opportunity of being able to "show" how Galileo could have proceeded. See in the appendix the "protocol" of the experiment, which shows that the absence of an instrument for measuring time may not have worried Galileo. This is a point discussed by historians of science, Alexandre Koyré even having claimed that Galileo only proposed experiments "on paper" that he never carried out.

3. Stengers, *Invention of Modern Science*, 89.

4. A different version of the same opposition: the sciences study phenomena and not the "essence of things."

5. On the importance of the "=" sign staged by the inclined plane, which constitutes the point where a "state of things" and its setting into a mathematical function are articulated, see Stengers, "Les affaires Galilée," 223–49.

6. Philosophers like to make the systematic doubt of Monsieur Descartes the thinking appropriate to the "age of the sciences." A deep misunderstanding (already denounced by Alexandre Koyré). Cartesian indubitability has nothing to do with the authority of the experimental fact any more than this authority has to do with the evidence of empiricists. Galileo refuses precisely that which the modern philosophers allow: that there must be something in their own experience that forms an ultimate, undoubtable reference. This divergence will become explicit with the controversy over "vital force" that raged until the middle of the eighteenth century and that pitted the Cartesians, for whom the conservation of movement is an evidence of reason, against those who accept the verdict of experimentation: if something is conserved, it is not mv, the quantity of movement, but mv^2 even if there is no reason that can justify a priori that velocity is squared. Reason must bend to the facts, not command them. It was Leibniz who initiated the controversy in 1686, by publishing "A Brief Demonstration of a Notable Error of Descartes and Others Concerning a Natural Law," in *Acta Eruditorum*; see Leibniz, *Philosophical Papers and Letters*, 296–302.

7. Latour, *Science in Action*, 153–55.

8. Latour, *Pandora's Hope*.

9. It was only when patents, previously restricted to "inventions," that is to say, to what responds to socio-industrial interests, were extended to what, for scientists, must be called "discoveries"—we have learned something new—that the calling into question by critics of the notion of discovery as a mere convention took place.

10. Needless to say, these obligations do not hold in the same way when scientists serve as experts, that is, intervene "beyond the boundary" in a way meant to be relevant for those who call them. We may deal them with "expertise battles" because there is no question of "succeeding" in giving to an "outside problem" the power to impose agreement on its interpreters.

11. The term *order-word* is at the same time both what responds to something self-evident, to what goes without saying, and thus is only spoken in the mode of redundancy, and what makes order reign, and so is spoken in mode of a reminder or warning. Those who would seek to distance themselves from it are immediately defined, not on the basis of the questions that they are able to ask but by this distance, and that is what will be the object of explanation. Here, the explanation generally refers to interests defined as "nonscientific" and provokes a reminder about the difference between science and nonscience.

12. Obviously this argument implies that one not confuse state apparatus and politics. The state allies itself with "Science" so as to limit the domain left open for politics, which it more or less puts up with.

13. This image of the public doesn't go back to the origins—Galileo sought an alliance with the public (Sagredo) against authority. It is, however, established in such a stable manner at the start of the twentieth century that it allows philosophers like Bachelard to identify science and an interest of the spirit (*esprit*—also, sometimes, mind), on the one hand, and opinion and a life interest, on the other, an identification that repeats all the dramaturgy of the sin/salvation duality. See, on this topic, Bensaude-Vincent, *La science contre l'opinion*.

14. Deleuze and Guattari, *A Thousand Plateaus*, 427. The "encasted" scientist corresponds with the image of the "immature" scientist, who has a childish approach to everything that isn't part of his science and whose childishness is, so the legend goes, the condition of his creativity.

15. On what Donna Haraway calls a "post-modern" becoming of science, see Haraway, *Modest_Witness@Second_Millenium*.

16. Apart from biotechnology, the massive undertaking around "nanotechnology" permits a glimpse of the probable future. Every distinction between the production of knowledge and the production of technological possibilities is triumphantly swept aside, and what unites researchers derives from the effect of training, rather than from the creation of demanding connections. A grandly visionary discourse circulates among researchers, industries, and states, and no one is capable, any longer, of distinguishing between projects and the rhetoric of the promise.

17. It is worth remembering that this book was published in 2006. Stengers was not thinking then of today's fact denialism. [TN]

Chapter 5. Troubling the Public Order

1. In Galileo's *Dialogue Concerning the Two Chief World Systems*. See previous chapter.

2. Dewey, *Public and Its Problems*, 116.

3. Dewey, *Public and Its Problems*, 184.

4. See chapter 3.

5. Dewey's position is analogous to that of Henri Bergson faced with "false problems." The best known of Bergson's examples is the problem of why there is something rather than nothing, which implies that one empties the scene of everything that exists, then considers the void as coming first, forgetting that it was necessary for us to create it.

6. Zask, "L'enquête sociale comme inter-objectivation," 149.

7. In what follows I will talk of "sociology" and of Dewey's failure, without any qualification. But what must be underlined is that this failure was not total, because there are some in the United States who are not without kinship with what he proposed, with the exception that most of them do not claim to be "scientists." On this subject, see Donzelot, Mével, and Wyvekens, *Faire société*. But I'm not ignoring that there are many ways of being a sociologist. My remarks are directed at "majoritarian" sociology, the kind for which scientificity is its gold standard. And if I relate it to Dewey's failure, it is because the latter also aimed at a "scientific" definition of the work of the sociologist.

8. This is how ethologists interpret the complicated character, which for the male is risky, of the rituals leading to copulation between

spiders: the male has to divert the female from being dominated by her experience "this is prey."

9. The case of "cold fusion," in which all the protagonists interested in what could be an inexhaustible source of cheap energy rushed in without waiting for the experimental verdict, constitutes for the experimenters a pathological case par excellence.

10. Latour, *Reassembling the Social*, 97–106.

11. The manner in which Harlow's systematic enterprise of torture was cited and taught for years says a great deal about the fact that "science that proves" can very easily become synonymous with abuse. Praise is due to Peter Singer, who was not only able to communicate the intolerable character of such "experiments" but also to show their "pseudoscientific" character. Experiments of this kind have since been proscribed. Also, they prompt no nostalgia: monkeys were tortured for the production of data with no interest (other than to have shown "scientifically" what we knew already). Singer, *Animal Liberation*.

12. Pierre Le Hir, "La société en mal de science," *Le Monde*, December 22, 2004.

13. This is the very idea of citizen juries or assemblies. When one reads Jacques Taminiaux's *Le théâtre des philosophes*, what comes to mind is an analogy between philosophers and the choruses of Greek tragedy, which Taminiaux rehabilitates, contra the scorn of philosophers. According to Taminiaux, tragedy was part of Athenian "political culture" because, through the mediation of the chorus, the Athenian public learned how to situate itself in relationship to tragic destiny, to have the experience of, the fear of, and the pity for the hubris that devours the hero without, for all that, allowing itself to be taken hostage by the insurmountable dilemmas that it provokes.

14. Dewey, *Reconstruction in Philosophy*, 27.

15. In fact, the brain in question is able to accept this imperative, because what is essential to it is that it ratifies what matters, which is the identification of science with "objective knowledge." In other words, it maintains the fable of origins, the fable that Galileo had constructed when announcing to his contemporaries the existence of a new kind of knowledge, one having the power to force a general redistribution of legitimacy, the power to demand a redefinition of "who has the right to talk about what." Even Heidegger, as I've already underlined, has been used, captured by some scientists, as an authoritative witness to the fact that their approach—a rather caricatural reductionism—is nothing other than the scientific approach as such, pursuing its irresistible process of conquest.

Intermezzo. The Creation of Concepts

1. Deleuze and Guattari, *What Is Philosophy?*, 22.

2. Deleuze and Guattari, *What Is Philosophy?*, 28.

3. Deleuze and Guattari, *What Is Philosophy?*, 41.

4. Deleuze and Guattari, *What Is Philosophy?*, 42.

Chapter 6. On the Same Plane?

1. Evidently what I am calling *testator* designates what today are called experts, but not all experts. There are notably some kinds of expertise that aim to elucidate (after a plane crash, for example). The specificity of the *testator* is that his questions actively, methodologically, entail the eventual elimination of what others designate as worthy of attention.

2. Unlike the "experimental proof," statistics allow only for a "showing" (*montrer*). They do not in the least produce a reliable interpretation of what is shown.

3. "Version" in Vinciane Despret's sense of the word. See Despret, *Our Emotional Makeup*.

4. Whitehead, *Science and the Modern World*, 207.

5. This is the big difference between magnetism, associated with cultivation of uncommon capacities, and hypnotism, which is experimental, subject to a protocol. See Méheust, *Le défi du magnétisme*, 533–49.

6. Along with Vinciane Despret and Didier Demorcy, I was able to pose this problem with regard to an experiment of this type, and we reproduced the manipulation [it was accomplished]. Experimental subjects did indeed exhibit the behavior that was anticipated, but when questioned afterward, all of them testified to what made them act, and the idea that their behavior was a "function" of the apparatus then became untenable: all of them had their own history, testifying to what it means to be subjected to an apparatus of this kind. The whole thing was the subject of a documentary, *The Vallins Experiment 1966–1999*, directed by Demorcy.

7. Deleuze and Guattari, *What Is Philosophy?*, 130.

8. See, notably, Haraway, *Modest_Witness@Second_Millenium*. Thanks to Maria Puig della Bellacasa for helping me to understand the difference between this art of the story and the "deconstructive" analyses of contemporary historian-sociologists.

9. Latour, *Making of Law*.

10. Popper, *Objective Knowledge*, 70.

11. When Popper talks about the amoeba, he opens up a category made to include those people against whom he didn't stop thinking, Marxists and psychoanalysts whom he reproached for making the defense of their hypotheses a life-or-death question.

12. See, in particular, Deleuze and Guattari, *A Thousand Plateaus*, 310–50.

13. I have a vivid memory of a news story on national radio that had emerged from a little village in rural France. Needing to take on an "office cleaner" and finding himself faced with a large number of equally well-qualified candidates, the mayor of this village selected someone at random. Scandal! He would have done better to be a bit creative than to ridicule the link between (s)election and merit, that is to say, the all-powerful reason that protects us from the arbitrary.

14. See Claverie, "Voir apparaître" and "La Vierge, le désordre, la critique."

15. See, notably, Toulmin, *Cosmopolis*, for a genealogical revisiting of the closed-door argument: it was to the Renaissance, which was troubled, violent, free, passionate, and torn apart by religious wars, that modernity said, "Not that ever again."

16. Stengers uses the word *bêtise*, which has no equivalent in English, especially in the use that Deleuze, whom she is following here, makes of it. The word *stupidity* derives from *stupor*, and it evokes the passivity of the one who will accept the most insane proposition. It could be adequate for the public rendered stupid by the way it is addressed when it accepts that before the laws of physics were established, people confused doors and windows. In French, *bêtise* is often associated with a touch of nastiness (*bête et méchant*). With Deleuze, it is active and predatory, actively dismembering what demands to be thought, giving authority to irrelevant questions, separating its victims from their capacity to object—rendering them stupid, in the other sense of the word, that is, "dumb." [TN]

17. It is the same with the phenomena that today are called "parapsychological," and there are analyses showing that the simple fact of taking an interest in them is a symptom.

18. On the functioning of the order-word, see Deleuze and Guattari, *A Thousand Plateaus*, 75–110.

19. Méheust, *Le choc des sciences psychiques*, 163–74.

20. Latour, "Guerre des mondes—Offre de paix," 61–80.

Chapter 7. We Are Not Alone in the World

1. Nathan used this affirmation as the title for a book published by Empêcheurs de penser en rond in Paris in 2001.

2. On the antifetishist critique, see Latour, *Cult of Factish Gods*, and *Pandora's Hope*, 266–91.

3. On this topic, see Souriau, "On the Work to Be Made."

4. Starhawk, *Dreaming the Dark*, 219.

5. It was through my contact with Léon Chertok that I understood the obscene character of this thinking of the winner who claims to inherit what he has destroyed. Chertok refused to accept that the Nazi destruction of Litwakia, as the Jewish people who were born there called their region, famous for the vitality of the Jewish culture, had a "meaning," was somehow a "sacrifice" necessary for some greater good. And Chertok extended this determined refusal to the destruction of the magnetists' tradition and then the caricaturizing of hypnosis. They did not pay the price for some tortuous progression toward the light! Chertok, Stengers, and Gille, *Mémoires d'un hérétique*.

6. This imposes on them a struggle on all fronts, a resistance as much to psychology as to the cult of the supernatural. So the "magic tricks" of conjurors themselves, who know how to direct the attention of and act on the perceptions of others, are not opposed to "real" magic. Conjurors are really practitioners of this art, as are those who communicate a knowledge of the mode that this knowledge requires, the mode that will transform whoever receives it. "To say the Goddess is reawakening may be an act of magical creation." Starhawk, *Truth or Dare*, 25.

7. What one must be wary of, in the reference to "other cultures" is poor taste. As Sybille de Pury underlines, those with whom djinns are concerned avoid uttering this word and systematically employ euphemisms. To include them in the title of a book would thus manifest an indifference on my part for what those they matter to fear. De Pury, *Comment dit-on dans ta langue?*

8. Claverie, "Voir apparaître," 167.

9. Deleuze and Guattari, *A Thousand Plateaus*, 348.

10. Deleuze and Guattari, *A Thousand Plateaus*, 350.

11. This wasn't necessary: the anecdote could have signified "evading an undertaking that requires the production of reproducibility." The fact that it came to signify "can only be of interest to 'unscientific'

or irrational minds," translates the poison of "Science": the anecdote seduces opinion, whereas the "true scientist" must hope that with the help of progress, the preciously anecdotal will be included by a science that is ever more powerful. It has to be said, in this instance, that this is an irrational hope, because it supposes that a question can be resolved without anyone taking an interest in it, that science can miraculously make intelligible what it has not only neglected but also despised.

12. See the study of Didier by Bertrand Méheust, *Un voyant prodigieux: Alexis Didier 1826–1866.*

13. Méheust, *Le choc des sciences psychiques*, 317–40.

14. See *Cosmopolitics* I. It is interesting to note here the manner in which, by making them shift from a problematic of existence to a problematic of knowledge, this operation of the "domestication of forces," in its reference to measurement, promoted the obscuring of the obligations of proof. Rationalist mechanists, such as Jean le Rond d'Alembert, dismissed any consideration that exceeded measurable facts as "metaphysical" and belonging in the store for useless accessories. *Force*, defined by measurable equivalence to its effect, then becomes perfectly legitimate, but at the cost of the call to measure, which will open up the field for the denunciation of a science that refuses to think and unilaterally "enframes." See also, on this topic, Stengers, *La guerre des sciences aura-t-elle lieu?*.

15. See Stengers, "Les affaires Galilée."

16. Nathan, *Nous ne sommes pas seuls au monde.*

Chapter 8. Ecology of Practices

1. Since the original publication of this book, the remains of the Kennewick Man have escaped the grasp of scientists. In June 2015, it was made public that scientists at the University of Copenhagen in Denmark determined through DNA from 8,500-year-old bones that the Kennewick Man is, in fact, related to modern American Indians. In September 2016, the US House of Representatives and Senate passed legislation to return the ancient bones to a coalition of Columbia Basin tribes for reburial according to their traditions. This was done on February 18, 2017, with two hundred members of five Columbia Basin tribes in attendance, at an undisclosed location in the area.

2. Competition can very easily accommodate weak results if the effect of their being announced is to attract investors, if others pay the price for consequences that have not been taken into account or were not

foreseen, or if—as is the case for medicine today—the price to pay is derisory in relation to the profit that a medicine whose harmful effects have been concealed makes it possible to rake in.

3. See Michel Tort's very important book, *Le désir froid*.

4. See chapter 4.

5. Such data can also be accumulated with general indifference, like the tons of bones that lie sleeping in museum storage, something that is condemned by the peoples whose ancestors have thus been defined as "data that might be useful in some future." Or they are accumulated in the expectation that a "genius" will one day create a relationship that will give them meaning. To be sure, they can also satisfy multiple forms of curiosity, but they can also become a poison, when satisfying this curiosity is assimilated to successful experimentation, when they are accorded the mode of existence that experimental facts are due.

6. Tort, *La fin du dogme paternal*.

7. Hence one sees professors of law recounting, and teaching their students, a pseudohistory in which the "law" progressively discovers an always more lucid, less "essentialist" identity, keeping carefully silent about the manner in which the evolution of law translates to the multiple histories that link it to political and social struggles, or to the dynamics of capitalist redefinition

8. See chapter 5.

9. In Leibniz, *Confessio Philosophi* (The Philosopher's Confession). See also Stengers, "The Betrayal of the Diplomats," in *Cosmopolitics*, 2:374–85.

10. The figure of the diplomat is presented in Stengers, "The Curse of Tolerance," in *Cosmopolitics*, 2:303–416.

11. There's nothing "supernatural" about this. For example, we know how attentive, critical, and alert people become when they take on the role of jury member in a court case.

Chapter 9. The Cosmopolitical Test

1. Originally published as seven separate books in 1997, the *Cosmopolitics* series was republished by La Découverte in two volumes in 2003. The two-volume English translation, published in 2010–11, reflects the organization of the 2003 publication.

2. In *Science and the Modern World*, Whitehead evokes the nemesis that waits upon "those who deliberately avoid avenues of knowledge"

(20), that is, those I am calling here "Galileo's inheritors," who avoid questions that do not let themselves be dismembered into how and why.

3. Needham, *Science and Civilisation in China*.

4. See Foucault, *Birth of Biopolitics*. These lectures, delivered in 1978–79, can be read as a veritable prophecy, one that would seem to have fallen on a rather large number of deaf ears, ears that were, moreover, equipped with tape recorders.

5. See Pignarre and Stengers, *Capitalist Sorcery*.

6. In his *Elective Affinities* (1809), Goethe made this opposition the central theme of a plot that contrasted amorous passion with the laws of marriage. See also Bensaude-Vincent and Stengers, *A History of Chemistry*.

7. Starhawk, *Truth or Dare*, 30–31, emphasis added. I notice that it's this same evocation that ends my *Hypnosis between Science and Magic*. I'll keep it here, nonetheless, because evocations have to be repeated.

Appendix. The First Experimental Apparatus?

1. From this hypothetical definition, Galileo can deduce that the heights descended (for the same incline of the plane) must, between them, be the squares of the time taken to fall those distances, but he is unable to measure precisely a descent time. However, his definition equally permits him to affirm that the descending times are, between them, as the velocities gained following this descent. And he can measure the relations between velocities gained for different heights. In effect, the velocity gained is the velocity with which the ball that has left the table continues its uniform horizontal movement while it is falling. Falling time being independent of this velocity, the latter is measured by the impact distance.

BIBLIOGRAPHY

Bensaude-Vincent, Bernadette. *La science contre l'opinion: Histoire d'un divorce*. Paris: Les Empêcheurs de penser en rond, 2003.

Bensaude-Vincent, Bernadette, and Isabelle Stengers. *100 mots pour commencer à penser les sciences*. Paris: Les Empêcheurs de penser en rond, 2003.

Bensaude-Vincent, Bernadette, and Isabelle Stengers. *A History of Chemistry*. Translated by Deborah van Dam. Cambridge, MA: Harvard University Press, 1996.

Chertok, Léon, and Isabelle Stengers. *A Critique of Psychoanalytic Reason*. Translated by Martha Noel Evans. Stanford, CA: Stanford University Press, 1993.

Chertok, Léon, Isabelle Stengers, and Didier Gille. *Mémoires d'un hérétique*. Paris: La Découverte, 1990. Republished under the title *Une vie de combats*. Paris: La Découverte/poche, 2020.

Claverie, Elisabeth. "La Vierge, le désordre, la critique: Les apparitions de la Vierge à l'âge de la science." *Terrain: Anthropologie et sciences humaines* 14 (1990): 60–75.

Claverie, Elisabeth. "Voir apparaître: Les 'événements' de Medjugorje." In *L'événement en perspective*, edited by Jean-Luc Petit, 157–76. Raisons pratiques 2. Paris: Éditions de l'École des Hautes Études en Sciences Sociales, 1991.

Collins, Harry, and Trevor Pinch. *The Golem: What You Should Know about Science*. 2nd ed. Cambridge: Cambridge University Press, 1998.

Deleuze, Gilles. *The Fold: Leibniz and the Baroque*. Translated by Tom Conley. Minneapolis: University of Minnesota Press, 1993.

Deleuze, Gilles, and Félix Guattari. *A Thousand Plateaus*. Translated by Brian Massumi. Minneapolis: University of Minnesota Press, 1987.

Deleuze, Gilles, and Félix Guattari. *What Is Philosophy?* Translated by Hugh Tomlinson and Graham Burchell. London: Verso, 1994.

de Pury, Sybille. *Comment dit-on dans ta langue? Pratiques ethnopsychi-atriques*. Paris: Empêcheurs de penser en rond, 2005.

Despret, Vinciane. *Our Emotional Makeup: Ethnopsychology and Self-hood*. Translated by Marjolijn de Jager. New York: Other Press, 2004.

Devereux, Georges. *From Anxiety to Method in the Behavioral Sciences*. Berlin: Walter de Gruyter, 1967.

Dewey, John. *The Public and Its Problems*. New York: Henry Holt, 1927.

Dewey, John. *Reconstruction in Philosophy*. New York: Henry Holt, 1920.

Donzelot, Jacques, Catherine Mével, and Anne Wyvekens. *Faire société: La politique de la ville aux Etats-Unis et en France*. Paris: Seuil, 2003.

Epstein, Steven. *Impure Science: AIDS, Activism and the Politics of Knowledge*. Oakland: University of California Press, 1996.

Foucault, Michel. *The Birth of Biopolitics: Lectures at the Collège de France, 1978–1979*. Edited by Frédéric Gros; translated by Graham Burchell. London: Palgrave-Macmillan, 2008.

Foucault, Michel. *The Hermeneutics of the Subject: Lectures at the Collège de France, 1981–1982*. Edited by Frédéric Gros; translated by Graham Burchell. London: Palgrave-Macmillan, 2005.

Ginzburg, Carlo. *Clues, Myths, and the Historical Method*. Translated by John and Anne Tedeschi. Baltimore, MD: Johns Hopkins University Press, 1989.

Gross, Paul R., and Norman Levitt. *Higher Superstition: The Academic Left and Its Quarrels with Science*. Baltimore, MD: Johns Hopkins University Press, 1994.

Haraway, Donna. *Modest_Witness@Second_Millenium. FemaleMan©_Meets_OncoMouse™*. New York: Routledge, 1997.

James, William. *The Writings of William James: A Comprehensive Edition*. Edited by John J. McDermott Chicago: University of Chicago Press, 1977.

Lapoujade, David. *William James: Empirisme et pragmatisme*. Paris: PUF, 1997.

Latour, Bruno. *The Cult of Factish Gods*. Translated by Catherine Porter and Heather MacLean. Durham, NC: Duke University Press, 2010.

Latour, Bruno. "Guerre des mondes—Offre de paix." *Ethnopsy—Les mondes contemporains de la guérison*, no. 4 (2002): 61–80.

Latour, Bruno. *The Making of Law: An Ethnography of the Conseil d'Etat*. Translated by Marina Brilman and Alain Pottage. Cambridge, UK: Polity, 2010.

Latour, Bruno. *Pandora's Hope: Essays on the Reality of Science Studies.* Cambridge, MA: Harvard University Press, 1999.

Latour, Bruno. *Reassembling the Social: An Introduction to Actor-Network Theory.* Oxford: Oxford University Press, 2005.

Latour, Bruno. *Science in Action. How to Follow Scientists and Engineers through Society.* Cambridge, MA: Harvard University Press, 1987.

Latour, Bruno. *We Have Never Been Modern.* Translated by Catherine Porter. Cambridge, MA: Harvard University Press, 1993.

Leibniz, Gottfried Wilhelm. *Confessio Philosophi: Papers Concerning the Problem of Evil, 1671–1678.* Translated by Robert C. Sleigh Jr. New Haven, CT: Yale University Press, 2005.

Leibniz, Gottfried Wilhelm. *Philosophical Papers and Letters.* Edited by L. Loemker. Dordrecht, Netherlands: Kluwer, 1975.

Lippmann, Walter. *The Phantom Public.* New York: Harcourt, Brace, 1925.

Méheust, Bertrand. *Le défi du magnétisme.* Vol. 1 of *Somnambules et médiumnité.* Paris: Les Empêcheurs de penser en rond, 1999.

Méheust, Bertrand. *Le choc des sciences psychiques.* Vol. 2 of *Somnambules et médiumnité.* Paris: Les Empêcheurs de penser en rond, 1999.

Méheust, Bertrand. *Un voyant prodigieux: Alexis Didier 1826–1866.* Paris: Les Empêcheurs de penser en rond, 2003.

Nathan, Tobie. *Nous ne sommes pas seuls au monde.* Paris: Les Empêcheurs de penser en rond, 2001.

Needham, Joseph. *Science and Civilisation in China.* 7 vols. Cambridge: Cambridge University Press 1954.

Perrin, Jean. *Atoms.* Translated by D. L. Hammick. New York: Van Nostrand, 1916.

Pestre, Dominique. *Science, argent et politique: Un essai d'interprétation.* Paris: INRA, 2003.

Pignarre, Philippe. *Qu'est-ce qu'un médicament?* Paris: La Découverte, 1997.

Pignarre, Philippe, and Isabelle Stengers. *Capitalist Sorcery: Breaking the Spell.* Translated by Andrew Goffey. London: Palgrave, 2011.

Pinch, Trevor. *Confronting Nature: The Sociology of Solar-Neutrino Detection.* Dordrecht, Netherlands: D. Reidel, 1986.

Poincaré, Henri. *The Value of Science.* Translated by George Bruce Halsted. New York: Science Press, 1907.

Popper, Karl. *Objective Knowledge: A Realist View of Logic, Physics, and History.* Oxford: Clarendon Press, 1966.

Rose, Hilary. "My Enemy's Enemy Is, Only Perhaps, My Friend." *Social Text* 46/47 (Spring/Summer 1996): 61–80.

Schatzki, Theodore, Karin Knorr Cetina, and Eike von Savigny, eds. *The Practice Turn in Contemporary Theory*. London: Routledge, 2001.

Serres, Michel. *Hermès III: La traduction*. Paris: Minuit, 1974.

Singer, Peter. *Animal Liberation: A New Ethics for Our Treatment of Animals*. New York: HarperCollins, 1975.

Sokal, Allan. "Transgressing the Boundaries: Toward a Transformative Hermeneutics of Quantum Gravity." *Social Text* 46/47 (Spring/Summer 1996): 217–52.

Souriau, Etienne. "On the Work to Be Made." In *The Different Modes of Existence*, 219–40. Translated by Erik Beranek and Tim Howles. Minneapolis, MN: Univocal, 2015.

Starhawk. *Dreaming the Dark*. Boston: Beacon Press, 1992.

Starhawk. *Truth or Dare*. San Francisco: HarperCollins, 1990.

Stengers, Isabelle. *Civiliser la modernité? Whitehead et les ruminations du sens commun*. Dijon: Les Presses du Réel, 2017.

Stengers, Isabelle. *Cosmopolitics*. 2 vols. Translated by Robert Bonnano. Minneapolis: University of Minnesota Press, 2010–11.

Stengers, Isabelle. *Hypnosis between Science and Magic*. Translated by April Knutson. London: Bloomsbury, 2022.

Stengers, Isabelle. *In Catastrophic Times: Resisting the Coming Barbarism*. Translated by Andrew Goffey. Lüneberg, Germany: OHP/Meson Press, 2015.

Stengers, Isabelle. *The Invention of Modern Science*. Translated by Daniel W. Smith. Minneapolis: University of Minnesota Press, 2000.

Stengers, Isabelle. *La guerre des sciences aura-t-elle lieu? Scientifiction*. Paris: Les Empêcheurs de penser en rond, 2001.

Stengers, Isabelle. "Les affaires Galilée." In *Eléments d'histoire des sciences*, edited by Michel Serres, 223–49. Paris: Bordas, 1989.

Stengers, Isabelle. *Réactiver le sens commun: Lecture de Whitehead en temps de debacle*. Paris: Les Empêcheurs de penser en rond, 2020.

Stengers, Isabelle. *Thinking with Whitehead: A Free and Wild Creation of Concepts*. Translated by Michael Chase. Cambridge, MA: Harvard University Press, 2011.

Stengers, Isabelle, and Bruno Latour. "The Sphinx of the Work." In Etienne Souriau, *The Different Modes of Existence*, 11–94. Translated by Erik Beranek and Tim Howles. Minneapolis, MN: University of Minnesota Press, 2015.

Taminiaux, Jacques. *Le théâtre des philosophes*. Grenoble: Jérôme Millon, 1998.

Tort, Michel. *La fin du dogme paternal*. Paris: Aubier, 2005.

Tort, Michel. *Le désir froid*. Paris: La Découverte, 1992.

Toulmin, Stephen. *Cosmopolis: The Hidden Agenda of Modernity.*
Chicago: University of Chicago Press, 1990.

Weinberg, Steve. *Dreams of a Final Theory.* London: Vintage, 1993.

Whitehead, Alfred North. *Science and the Modern World.* New York,
Free Press, 1967.

Zask, Joëlle. "L'enquête sociale comme inter-objectivation." In *La croy-
ance et l'enquête: Aux sources du pragmatisme*, edited by Bruno
Karsenti and Louis Queré, 141–63. Paris: Éditions de l'École des
Hautes Études en Sciences Sociales, 2004.

INDEX